Intelligent Positioning
GIS-GPS Unification

GEORGE TAYLOR

Professor and Research Leader
School of Computing, University of Glamorgan
Wales, UK

and

GEOFF BLEWITT

Research Professor
Nevada Bureau of Mines and Geology/Seismological Laboratory
University of Nevada, Reno, USA

John Wiley & Sons, Ltd

Other Wiley Editorial Offices

John Wiley & Sons Inc., 111 River Street, Hoboken, NJ 07030, USA

Jossey-Bass, 989 Market Street, San Francisco, CA 94103-1741, USA

Wiley-VCH Verlag GmbH, Boschstr. 12, D-69469 Weinheim, Germany

John Wiley & Sons Australia Ltd, 42 McDougall Street, Milton, Queensland 4064, Australia

John Wiley & Sons (Asia) Pte Ltd, 2 Clementi Loop #02-01, Jin Xing Distripark, Singapore 129809

John Wiley & Sons Canada Ltd, 22 Worcester Road, Etobicoke, Ontario, Canada M9W 1L1

Wiley also publishes its books in a variety of electronic formats. Some content that appears in print may not be
available in electronic books.

Library of Congress Cataloging-in-Publication Data
Taylor, George.
    Intelligent positioning : GIS-GPS unification / George Taylor and Geoff Blewitt.
        p.   cm.
    Includes index.
    ISBN-13: 978-0-470-85003-9
    ISBN-10: 0-470-85003-5
    1. Global Positioning System.   2. Geographic information systems.   I. Blewitt, Geoff.   II. Title.
    G109.5.T35 2006
    910.285 – dc22

                                                                                         2005029799

British Library Cataloguing in Publication Data

A catalogue record for this book is available from the British Library

ISBN-13 978-0-470-85003-9 (Hardback)
ISBN-10 0-470-85003-5 (Hardback)

Typeset in 10/12 pt Times by SNP Best-set Typesetter Ltd., Hong Kong
Printed and bound in Great Britain by TJ International, Padstow, Cornwall
This book is printed on acid-free paper responsibly manufactured from sustainable forestry in which at least two
trees are planted for each one used for paper production.

Contents

Foreword

The over-riding lesson that you will learn from this excellent book is that there is positioning, and there is *intelligent* positioning. As more and more people use and exploit GNSS signals to acquire position (whether surveyors or drivers), it has become clear that the true measure of a position is the context in which it can be set. Surveyors using GPS need to understand how the signal is generated, transmitted and decoded so that they can make informed judgements about the level of error that they are dealing with. The discussion of GPS error indicators in this book, such as dilution of precision (DOP), is the most lucid I have ever read, and gives surveyors and others authoritative information on the absolute performance of a positioning service.

Drivers, at the other extreme from surveyors, also need context for their positioning service. To plot GPS positions on a map requires a good understanding of maps, projections and geodetic datums. All of these subject areas are discussed here in an accessible style, with links to online resources and through practical case studies. What is written here will allow drivers and others to be informed users of GNSS, for example, to know why and how gross errors might be generated and when to look out for them.

I'd like to congratulate the authors on writing so clearly on topics of great complexity, and to recommend this book to all with a desire to truly understand intelligent positioning.

Jonathan Raper
Head of Department and Professor of Geographic Information Science
Department of Information Science, City University, London, UK
November 2005

Preface

Both geodetic and geographical science have benefited greatly from the technological revolution that has recently dominated human achievement, more specifically, from satellites for Earth observation and communication, combined with computers with an ever-accelerating rate of improved data processing power and capacity. Satellite positioning and computerized spatial data handling have fundamentally transformed the way in which many areas of research and commercial activity are conducted. The combination of GPS and GIS is used for a broad spectrum of investigative experiments ranging from the study of the Earth's plate tectonics to epidemiological research into the incidence and distribution of children's respiratory diseases. Furthermore, established applications, such as vehicle navigation and emergency services decision support, all now combine GIS and GPS in order to provide a solution.

The complexity and depth of this integration of GIS and GPS vary considerably. Many procedural applications, such as crime pattern analysis, may use GPS to provide accurate positioning of an incident together with GIS to record and analyse the distribution and occurrence of crime. This represents a combination of the two technologies, but they are not really being integrated at the lowest level. At the other extreme, a vehicle-tracking system may process data directly from a GPS receiver together with constraints imposed by digital maps to determine the most statistically likely position.

This book details the research and development activities undertaken by the two authors. These include: the augmentation of GPS-derived positioning using Digital Terrain Models (DTM) and digital mapping, embedded GIS in Intelligent Transport Systems, GIS and GPS in mobile phone positioning, and GIS and GPS in real-time location-based tourist information systems. This book also provides a brief overview of GIS, GPS and datum transformations and projections. One chapter provides an overview of current commercial applications that use the combined technologies: one focuses on applications that use superficial integration, while the other looks at applications with deep integration.

The intended audience for this book includes graduate students in geomatics and civil engineering, geomatics researchers, and R&D professionals in industry. Some chapters may be useful for undergraduate courses in geomatics and civil engineering. To maintain relevance as technology continues to progress, we have chosen to focus on fundamental principles and concepts rather than specific examples of off-the-shelf technology. Therefore, this is in no way a guidebook to commercially available systems, nor are specific systems even mentioned. Rather, the goal is for you, the reader, to appreciate our approach to the integration of these two technologies at a low level, and be able to apply this kind of approach to your own specific applications. So even if the specific algorithms described here become obsolete (as in time they always do), the general approach will still be relevant.

Acknowledgements

A number of chapters in this book are based on research projects undertaken by the authors and their researchers, including PhD students. We would like to acknowledge the contributions to previously published outputs of these projects, which have contributed to the work described here.

Contributions to original papers and conference presentations: David Kidner, Chris Brunsdon, Andrew Olden, Dörte Steup, Marylin Winter, Jing Li, Simon Corbett, Adrianna Car, Emma-Jane Mantle, Marion Garten, Phil White, Kenny Steele, and Stuart Cole.

List of Abbreviations

2D	two-dimensional
3D	three-dimensional
AM/FM	Automated Mapping/Facilities Management
ANN	Artificial Neural Networks
A/S	Anti-Spoofing (GPS signal encryption)
ATI	Applied Technology Institute
C/A	Course/Acquisition (GPS code)
CAD	computer-aided design
CDOP	Correction DOP
CLDS	Canada Land Data Systems
CSREES	Cooperative Research, Education and Extension Service
DB	database
DBMS	DB management System
DEM	Digital Elevation Model
DGPS	Differential GPS
DOD	Department of Defense (US)
DOE	Department of the Environment (UK)
DOP	Dilution of Precision
DORIS	Doppler Orbitography and Radiopositioning Integrated by Satellite
DLL	Dynamic Link Library
DR	Dead Reckoning
DTM	Digital Terrain Model
DRI	Drive Restriction Information
EKF	Extended Kalman Filter
ESRI	Environmental Systems Research Institute
FIFO	first in first out
GIPSY	GPS-Inferred Positioning System
GIS	Geographical Information System
GLONASS	Global Orbiting Navigation Satellite System (Russia)
GML	Geography Mark-up Language
GNSS	Global Navigation Satellite System
GPS	Global Positioning System (US)
GUI	Graphical User Interface

HDOP	Horizontal DOP
HSGPS	High Sensitivity GPS
IERS	International Earth Rotation and Reference Frame Service
IGS	International GNSS Service
IMP	Initial Matching Process
INS	Inertial Navigation Systems
InSAR	Interferometric Synthetic Aperture Radar
IRRF	Intelligent RRF
ITRF	International Terrestrial Reference Frame
ITRS	International Terrestrial Reference System
ITS	Intelligent Transport Systems
L1 and L2	carrier frequencies (GPS)
LBS	Location-Based Services
LIDAR	Light Detection and Ranging
LIFO	last in first out
LIS	Land Information System
MDOP	Mapping DOP
MMGPS	Map-Matched GPS
MSL	mean sea level
NAD	North American Datum
NGS	National Geodetic Survey (US)
NRCan	Natural Resources Canada
NSF	National Science Foundation (US)
OGC	Open GIS Consortium
OMMGPS	Odometer-derived MMGPS
OO	Object-Oriented
OS	Ordnance Survey (UK)
OSGB	Ordnance Survey Great Britain
P	Precise (GPS code)
PARAMOUNT	Public Safety & Commercial Info Mobility Applications & Services in the Mountains
PDOP	Position DOP
PRN	Pseudo-Random Number (GPS code)
PULSAR	Princeton University's Large-Scale Automobile Routing
RAIM	Receiver Autonomous Integrity Monitoring
RDBMS	Relational DBMS
RDBS	Relational DB Structure
RMSE	Root Mean Square Error
RRF	Road Reduction filter
RTK	Real Time Kinematic
S/A	Selective Availability (intentional GPS signal error)
SAR	Search-And-Rescue
SDSS	Spatial Decision Support System
SMP	Subsequent Matching Process
SMS	Short Messaging Service

SQRT	Square Root
SLR	Satellite Laser Ranging
SPS or SPCS	State Plane Coordinate System
SVN	Satellite Vehicle Number
TCN	Terrain Contour Navigation
TIGRIS	Topologically Integrated Geographic Resource Information System
TIN	Triangulated Irregular Networks
TRN	Terrain-Referenced Navigation
UTM	Universal Transverse Mercator
XOR	Exclusive OR (binary function)
UK	United Kingdom
US	United States
VDGPS	Virtual Differential GPS
VDOP	Vertical DOP
VDU	Visual Display Unit
VLBI	Very Long Baseline Interferometry
VR	Virtual Reality
W3C	WWW Consortium
WGS	World Geodetic System
WWW	World Wide Web
XML	Extensible Mark-up Language

Introduction

1. Do You Really Know Where You Are?

The Global Positioning System (GPS) is a system of satellites that broadcast radio signals, allowing you, 'the user', to find out where you are. To do this requires a 'GPS receiver'. This might be a hand-held device, or a system mounted in your car, boat or plane.

Let's think about what it really means 'to know where you are'. What you might find on the screen of a GPS receiver are some numbers that tell you your longitude, latitude and perhaps your height. Given these numbers every few minutes, you could plot a course on a map and see where you are, and where you are heading. This all sounds very simple.

The problem is not that simple. Is the map going to tell you what you really want to know? Suppose you are driving, and want to know where to turn next if you are heading towards a certain address. You might have a street map, but does the map show a grid of longitudes and latitudes? Even if it did, are you sure it is the same type of longitude and latitude as used by GPS? Every country has a different system of coordinates – does GPS know this? Not to mention, there is the practical difficulty of plotting your course, on a map while driving!

Of course, most of you are well aware that you can buy cars today that are equipped with a Global Navigation Satellite System (GNSS), such as GPS, which does all this work for you. Well, it's not GPS that is telling where to turn next, or which street you are on. The only thing GPS does is report your longitude, latitude, and height within an international coordinate system known as WGS-84 (we'll come back to that later). Rather, it is a geographical information system (GIS), a sophisticated computer program, user interface and display, which uses the GPS coordinates as an input. The GIS processes a sequence of your GPS coordinates and can give you the type of navigational information that is truly useful to you.

This is one of the most obvious examples of how GPS and GIS have become integrated. To the car driver, it might appear to be so well integrated that it is

Intelligent Positioning: GIS-GPS Unification G. Taylor and G. Blewitt
© 2006 John Wiley & Sons, Ltd

just one system, which the driver simply calls 'my GPS system'. It would be more correct to call it a GPS-GIS system.

2.   How Active Is Your Map?

In the above example, is it simply the case that GPS is handing over some coordinates for GIS to plot on a map? That is perhaps one way of looking at it, but it is not a particularly useful way if you want to understand how deeply integrated GPS-GIS systems work. Deeply integrated GPS-GIS systems treat both the GPS data and digital map data as information that can be used to provide you with your most statistically likely position, and much more.

As we shall see, this is not simply a matter of plotting the GPS position onto a digital map. The map itself provides important data that is used mathematically as part of the 'inversion process', which produces a best estimate of your location. In this view, the map is not a passive slate upon which positions are plotted, but rather it defines the space of possible user locations, depending on the application. For example, car locations can be restricted to known streets. There may also be rules that disallow a car's most 'likely position' to be on one-way streets if the car is not heading in the appropriate direction.

As another example, the car's height above sea level might be restricted so that it must lie on the Earth's surface. This rule projects the three-dimensional space of GPS onto a two-dimensional space of acceptable answers. Through the mechanism of least squares estimation, this 'rule' can be equivalently considered as a set of data, though of course such 'data' are not actual measurements, but are rather the data provided by a digital terrain map.

3.   Levels of GPS-GIS Integration

At the loosest level of integration, the GIS could simply take the GPS reported results and display them on a map. Here the map need be nothing more than an electronic 'photo-copy' (raster scan) of a hand-drawn map. There's nothing very intelligent about this type of integration. It's simply a matter of making the results look presentable.

Loosely integrated GPS-GIS represents perhaps the type of system most people would consider when thinking about GPS. Even the least expensive hand-held receivers on the market today have some basic GIS functions built into them. For example, the GIS part of the system can convert the WGS-84 coordinates produced by the GPS part of the system, and convert them into a set of map coordinates that are appropriate for that country or state. The 'GPS receiver' display might then show your position superimposed on a digital map of the country, with user-selectable features such as magnification (zoom), and choice of what is shown on the map (for example, roads, rivers, landmarks, boundaries, etc.).

At a slightly deeper (but still loose) level of integration, GPS receivers often allow the user to stop at a certain location (a 'way point') and enter what type of feature might be present, for example, a road junction. In this way, the GPS-GIS system allows users to create a GIS database, which can then be used to create a custom-built map. Even more deeply integrated GPS-GIS systems can then navigate the user (tell the user in which direction to move) in order to find a previously recorded feature that now is included on the digital map.

In the most deeply integrated, 'intelligent' GPS-GIS, digital maps are not simply for output purposes to display the result (although this is certainly an essential function), but are themselves a source of valuable information that can be used to improve the accuracy of the reported position. This information can equivalently be thought of as a 'space', as a 'set of rules' or 'constraints', or simply as 'data'. This flexibility in the way we can think about digital map data allows different ways for this information to be applied. Thus, in some situations, the application of logical rules may be the most effective means of applying the information, and in other situations it may be most convenient to treat the map as a set of pseudo-measurements. In this book, examples are given where the reader might recall how they relate to the various active ways in which digital map information can be brought to bear on the navigation problem.

4.   Overview of the Book

Chapter 1 provides a short introduction to basic geo-spatial concepts that under-pin geographical information systems. Details include the definition of a GIS (what and why), its terminology, and a brief discussion on modeling the real world in geometrical and descriptive (attribute) terms. The discussion is mainly about the technology with a very brief outline of application areas.

Chapter 2 introduces the principles of GPS theory, and provides the back-ground for later chapters. The theoretical treatment has been simplified to provide a starting point for a mathematically literate user of GPS who wants to understand how GPS works, and to get a basic grasp of GPS theory and terminology.

Chapter 3 describes the nature of the Earth in relation to mass and gravity. It discusses the relationship between orthometric heights and GPS heights with the requirement, and related procedures, to transform GPS height to mean sea level (MSL) height, such as Ordnance Datum Newlyn height. A brief introduction is given to projections, which are required to transfer measurements and positions from the ellipsoid onto a flat surface suitable for making into a map.

Chapter 4 reports on current commercial applications that integrate GIS and GPS. There are many applications that loosely integrate GIS-GPS. However, the emphasis is on current industrial applications that use embedded or deeply integrated GIS and GPS. These applications include intelligent transport systems (ITS), and various location-based services (LBS).

Chapter 5 describes the Road Reduction Filter (RRF) in detail. RRF is a method of detecting the correct road on which a vehicle is travelling. In the work described here, the position error vector is estimated in a formal least squares procedure, as the vehicle is moving. This estimate is a map-matched correction that provides an autonomous alternative to differential GPS (DGPS) for in-car navigation and fleet management.

Chapter 6 focuses on our patented method for the reduction of errors in raw GPS position data, based on the known road geometry (road shape), as predefined by a digital map. One problem with the approach presented in Chapter 5 is that, for example, if the GPS satellites happen to have selective availability (S/A) switched on, the position's output from a GPS receiver can have errors of up to 100 m, so, while it may be possible at a particular point in time to identify the correct road, the position along the road may be in error by up to 100 m, and is frequently plus or minus 20 m, as shown by the 'along-track' error (see Figure 5.13). This along-track error cannot be resolved for a straight road, but it can be resolved if the road changes direction, or if the vehicle turns a corner. Quality measures can be derived and used to place confidence bounds for rigorous decision-making on the reliability of such an error model. A formula is derived for the quantity "mapping dilution of precision" (MDOP), defined as the theoretical ratio of position precision using map-matched corrections to that using perfect DGPS correction. This is shown to be purely a function of route geometry, and is computed for examples of basic road shapes. MDOP is shown to be favourable unless the route has less than a few degrees curvature for several kilometres. MDOP can thus provide a vehicle driver with an objective estimate of positioning precision. Precision estimates using MDOP are shown to agree well with 'true' positioning errors determined using high precision (cm) GPS carrier phase techniques.

Furthermore the use of digital road map intelligence will be included in the RRF to improve the efficiency of isolating the correct road centre-line, e.g. one-way traffic direction, direction of travel at a roundabout, or along a dual carriageway.

Chapter 8 describes the further development of the test-bed application, called map-matched GPS (MMGPS), and described in Chapters 5 and 6. This method uses absolute GPS positioning, map matched, to locate the vehicle on a road centre-line, when GPS is known to be sufficiently accurate. MMGPS software has now been adapted to incorporate positioning based on odometer-derived distances (OMMGPS), when GPS positions are not available. Relative GPS positions are used to calibrate the odometer.

Finally, we'd like to point out to the reader that a list of abbreviations has been provided to define the many abbreviations found in this book. Whereas abbreviations are typically defined on first usage (and sometimes more), we understand how difficult it is to remember unfamiliar abbreviations when there are so many of them involved in high technology such as GPS and GIS!

1

GIS: An Overview

This chapter provides a short introduction to basic geo-spatial concepts that underpin geographical information systems (GIS), i.e. data, its collection and input, storage and organization, processing and analysis, reporting and output. Details include the definition of a GIS (what and why), its terminology, a brief discussion on modelling the real world in geometrical and descriptive (attribute) terms, raster and vector models and associated spatial operations. The focus will be on the technology with very brief outline on application areas. Many references are provided for more in-depth information on GIS technology.

1. Introduction

Information is derived through the processes of collecting, collating, structuring, and analysing factual data. Geographical information is a term which is used to encompass many forms of information derived from many different data sources:

> Geographical Information is information which can be related to specific locations on the Earth. It covers an enormous range, including the distribution of natural resources, descriptions of infrastructure, patterns of land use and the health, wealth, employment, housing and voting habits of people.
>
> [DOE, 1987]

Geographical information systems (GIS) are now as widely used as many other desktop information handling and processing tools, such as database management systems and spreadsheets. GIS share the same Windows-style graphical user interfaces (GUI) as other desktop packages, (see Figure 1.1, Plate 1). The fundamental technology requirements of spatial data-based applications are now well understood and addressed by the majority of GIS vendors and researchers. A wide range of disciplines use GIS in their daily activities, for example, any

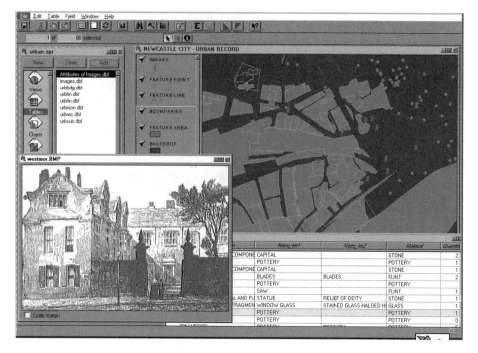

Figure 1.1   *GIS interface*

driver using an in-car navigation system or any Internet user/surfer geographically locating an address by UK postcode or a US zip code, using any of the popular World Wide Web (WWW) sites for mapping such as Streetmap® (http://www.streetmap.com/) or MapQuest® (http://www.mapquest.com/). GIS has become the overall name for a continuum of systems ranging from those that simply display maps for visualization purposes, including computer software packages such as computer-aided design (CAD) systems, to those that provide very powerful spatial and statistical analysis of spatial and related data and those providing full immersive virtual environments complete with haptic interface, devices that allow human–machine interaction through force and touch. The proceedings are available of the series of symposiums on 'Haptic Interfaces for Virtual Environment and Teleoperator Systems', now in its tenth year [IEEE-VR2002, 2002].

There are now many good books that cover, in great detail, the basic concepts of GIS, Bernhardsen [1992], Jones [1997], Burrough and McDonnell [1998] Demers [2000], and Longley *et al.* [2001], to name but a few. A good reference for many introductory texts in GIS is provided on the National GIS/GPS Integration Team web pages [Information Technology Education, 2005]. One of the first comprehensive texts on the principles of GIS was written by Burrough (1986); this is still a seminal work covering the subject.

2. GIS

The definitions used to describe a geographical information system apply equally to a land information system (LIS) and an automated mapping/facilities management system (AM/FM) or any other spatially referenced information management systems. What is a GIS? Here's one definition: 'a set of tools for collecting, storing, retrieving at will, transforming, and displaying spatial data from the real world for a particular set of purposes' [Burrough, 1986]. There is, however, still no widespread agreement on the definitions and taxonomy of these systems. There are no hard boundaries between their use, application or specification. This is particularly true at the international level. The United Kingdom and the United States generally use GIS to cover all types of systems. The Lands Directorate, Environment Canada, original users of Canada Land Data Systems (CLDS) [Tomlinson, 1984], the first recognized GIS in the world [Crain and MacDonald, 1983], provide a short and clear definition: 'Geographical Information Systems (GIS) are systems designed to store and manipulate data which relate to locations on the Earth's surface' [Poiker *et al.*, 1985]. These early definitions of what constitutes a GIS are still valid today, and are intrinsic to a more modern and broader definition such as: 'Any system that handles geo-referenced data may be considered to be a GIS.'

2.1. The Basic Idea

Geographical data is most often separated into two components: spatial data and attribute data. Spatial data is used in the visualization and manipulation of real-world objects in a computer model, e.g. roads, buildings, crime locations. Typically, spatial data is presented as features on a digital map. Attribute data (textual, numeric, photographic, etc.) describes these real-world objects, e.g. name, cost, size, and an image of what the object looks like. These two components often are stored in different data structures, in separate databases. Many commercial and research systems developed to use geographical data utilize this method, typical examples are APIC4 GIS [APIC, 2003] and Environmental Systems Research Institute ARCGIS/INFO, possibly the most widely used GIS in the world. Almost all commercial systems store and manipulate all data, both spatial and attribute, stored in their own proprietary database and file structures. One exception to this is the use of an external database, for attribute data only, such as Microsoft Access or the ArcSDE/Oracle database to store both spatial and attribute data. The division of data into two data models may provide a more efficient operating environment for graphical display, editing and enquiry, which will also provide a comprehensive attribute retrieval and analysis system. The attribute database is almost universally based on the relational database structure; however, object-oriented (OO) databases have been used in some systems, e.g., GE Smallworld [GE Energy, 2004].

GIS data is usually organized into themes, also called layers, overlays, or coverages. These can be conceptualized as overlaid maps (Figure 1.2). Each theme

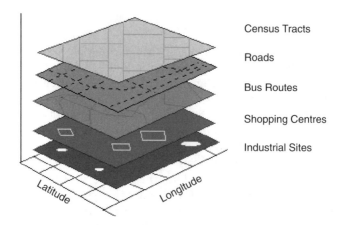

Census Tracts

Roads

Bus Routes

Shopping Centres

Industrial Sites

Figure 1.2   *GIS themes or layers*

contains geographical features with similar qualities and characteristics, e.g. one layer may contain only roads, another water features (rivers, lakes, etc.), land use zones, health data, soil type, rainfall, etc.

3.   Functionality

A computerized GIS has numerous and varied capabilities. However, many definitions reference the basic components of input, storage, analysis and output.

3.1.   Input

The sources of spatial data for a GIS encompass all current survey techniques. These include field survey, photogrammetry, remote sensing, GPS, digitization, digital photography, LIDAR (airborne 'light detection and ranging'), etc. All these methods now produce coordinate referenced data in an acceptable digital form for input into a GIS, although this input may demand a translation process. Moreover, existing paper mapping still provides a major source of spatial data. Manual digitizing is slow, expensive and error-prone. Automated digitizing systems have now reached a level of sophistication acceptable to many users. Fairbairn and Taylor [2002] discuss in detail data collection methods and issues in *Virtual Reality in Geography*. They present an assessment of available data collection options and the use of each method, with illustrative examples showing the wide range of possible solutions.

3.2.   Storage

The efficient storage, management and retrieval of large amounts of textual data are a requirement of almost all information systems, including GIS. A database management system (DBMS) is the software tool for this task. The relational database management system (RDBMS) is now the accepted industry standard to accomplish this. Object-oriented database management systems (OODBMS) are now of great interest for information system implementation in general and a few are available commercially. However, OODBMS do not have formal foundations, unlike the relational model, and no common data model. Some GIS do offer some form of object-oriented database model, but all fall well short of the true OO paradigm of complex objects, object identity, encapsulation, types or classes and inheritance. Smallworld GIS, now part of GE Network Solutions, is a well-established example which also provides its own object-oriented language: Smallworld Magik™ [Smallworld, 2003]. The principles of the relational structure were first presented by Codd [1970], who realized that the discipline and rigour of mathematics could be used in the field of database management. The relational structure is based on the mathematical set theory of relations and predicate calculus. A comprehensive and rigorous discussion of database models and structures is given by Date [2000].

3.3. Analysis

The database structure and spatial data format used will always affect the types of algorithms available for spatial analysis. Conversely, the type of analysis necessary may require a certain data format or database structure. For example, the analysis of utility or transport networks is really only feasible with vector data format. An example vector land parcel theme is shown in Figure 1.3 (Plate 2). Raster and vector data are introduced later in this chapter.

The analytical functions required to operate on the spatial and non-spatial elements of geographical data range from those that retrieve simple subsets of information for display to those that use spatial relationships and topological overlay to create new objects. The requirement for spatial analysis varies considerably with the application, e.g. a LIS used to maintain a fiscal cadastre will require very little, but a multi-criteria spatial decision support system (SDSS) used for hydrological modelling for flood prediction will require lots.

The differences in the raster and vector representation of location information are significant in terms of strategies for implementation. Certain operations will be easier in one form than in the other. However, a set of fundamental analytical techniques may be defined that will apply to all representations of spatial data. This is presented below, taken from Berry (1987). Another useful categorization of general types of spatial analysis is provided by Longley *et al.* (2001).

Reclassifying Maps. This may be considered the most fundamental type of analytical operations. A new map is created by assigning thematic values to the categories of an existing map (overlay). These values may be assigned as a function of the initial value, its position, contiguity, size or shape of the spatial configuration of the individual categories. For example, a map of soil types might be

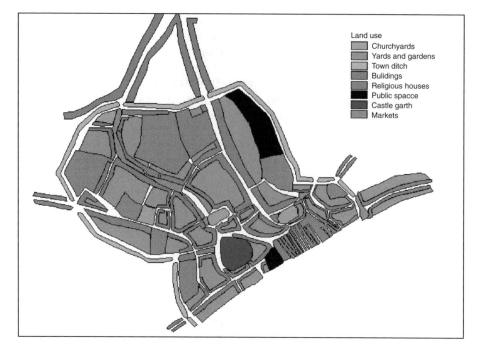

Figure 1.3   *Vector land use map*

assigned values that indicate the relative suitability of each type for residential development, and values for levels of suitability for plant growth.

Overlaying Operations. These operations involve the creation of a new map on which the value assigned to every point or region (polygon) is a function of the independent values associated with that location on two or more existing overlays. Location-specific overlaying will, for example, identify particular combinations of values, e.g. land use and slope. Most typical of environmental analysis is the manipulation of quantitative values, using arithmetic operators, e.g. given two overlays of assessed land values for 1980 and 1990; a map may be generated showing the percentage change in land values over that 10-year period.

  Measuring Distance and Connectivity. This operation involves computing optimum routes in units such as length, time or cost. This may result in the actual distance between a pair of locations or the identification of equidistant zones around a location or set of locations. For example, the route by road between two cities, zones of equal travel time from a given location. Another distance-related class operation is concerned with the nature of connectivity between locations. This may involve establishing the intervisibility between locations or the analysis of a public utility pipe network, to determine demand and capacity values.

Characterizing Neighbourhoods. This final group of operations includes procedures that create a new map where the value assigned to a location is computed as a function of independent values surrounding that location. A neighbourhood

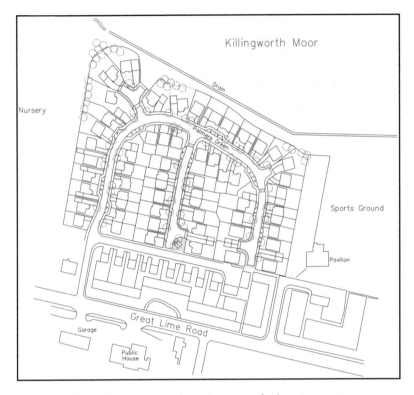

Figure 1.4 *Large-scale vector map of a housing estate*

may be a uniform geometric shape, such as a circle or square, or perhaps 10 km 'down-wind'. One example of this operation is the calculation of slope from elevation values. Another example is the summarizing of thematic values, e.g. maximum income levels, diversity of vegetation types. Figure 1.5 displays LIDAR data characterized by height.

3.4.   Output

Output from a GIS may be a standard cartographic product such as a map sheet or a simple table of data entries. Alternatively, it may be the result of a query or some analysis. Modern technology will produce high quality displays of maps, tables, graphs, reports, etc. on VDU, printer and plotter. Output may be recorded on magnetic media for remote use or for further analysis by other systems.

4.   Fundamental Concepts

4.1.   Features

Geographical information, sometimes referred to as spatially related information in order to differentiate it from purely topographical information, is most

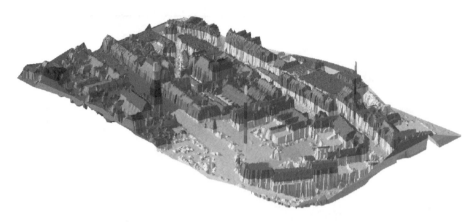

Figure 1.5   *LIDAR data, categorized by height*

often represented in the form of features or objects. A feature is an item of direct interest and may be defined as a number of spatial and aspatial elements, treated as an indivisible set, e.g. a land parcel boundary, a gas pipeline or a tree. Each feature in a particular data set will have a unique reference. When features are displayed graphically, different line types, symbols and colours are often used to enhance their visual classification.

4.2.   Spatial Elements

Spatial elements, or spatial entities, are the basic building blocks of spatial data. The most commonly defined spatial elements are points, lines and areas. These elements are used to represent geographical information on a map. A feature is constructed from one type of element but may consist of any combination of these elements.

4.3.   Attribute Information

Non-spatial attributes, sometimes referred to as aspatial data, are those properties of one or more spatial entities that are used to graphically represent a feature. 'An attribute is a property of an entity, usually used to refer to a non-spatial qualification of a spatially referenced entity' [DOE, 1987]. Figure 1.6 displays an area feature with associated attributes. The attributes for a land parcel feature could include ownership, freehold and leasehold details, or for a road feature they could be width, name, kind, etc.

5.   Spatial and Geographical Data

The term 'spatial data' applies to any data concerning phenomenon which is referenced by location in multidimensional space. Geographical data, more specifically, is spatial data pertaining to the Earth [Peuquet, 1984].

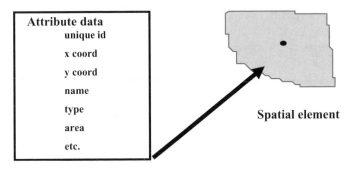

Figure 1.6 *Area feature with associated attributes*

5.1. Spatial Referencing

The power of GIS data handling is in its spatial or geo-referencing functionality. Direct spatial referencing uses some form of spatial coordinate system to locate data items. This coordinate system is most likely to be geographical coordinates in latitude and longitude or planar grid coordinates in a national grid system, e.g. the State Plane Coordinate System (alternatively abbreviated as SPS or SPCS) used in the United States, Puerto Rico and the US Virgin Islands, [King, 2004], the OS National Grid in the UK [Ordnance Survey, 2004], latitude and longitude or rectangular coordinates derived from a Transverse Mercator projection [Harley, 1975].

Discrete methods of spatial referencing use discrete units on the Earth's surface, e.g. country, county, postcode. Many of these methods are indirect. Indirect means that the method provides a key or index, which can be used with a table to determine latitude/longitude coordinates, for example, a post code is an indirect spatial reference. Rather than give national grid coordinates for a place directly, it provides a unique number which can be looked up in a table which provides a spatial reference using coordinates on a map. Because these methods are indirect, it is important to consider the precision of these systems. Precision is related directly to the size of the discrete unit which forms the basis of the spatial referencing system. Many methods of indirect or discrete spatial referencing are in common use. The precision of street addresses varies as spatial references vary, it is highest for houses in cities and is lowest for rural addresses or post office box numbers, where the address may indicate only that the place is somewhere in the area served by the post office. Postcode systems have been set up in many countries; two well-known systems are the US zip code and the UK postcode. The US zip codes are designed to assist with mail sorting and delivery; the codes are hierarchically nested, states are uniquely identified by one or more sets of the first two numbers, a 5-digit zip code identifies the area served by a single post office, which gives precision of many city blocks. The 9-digit zip potentially provides a much higher level of spatial resolution, but problems exist, e.g.

buildings may have different codes for different floors, zip code boundaries can overlap and may be fragmented. In the UK postcode system there are about 1.6 million postcodes covering approximately 24 million addresses in the UK. This system is mixture of numeric and alphabetic characters, unlike most other postcode systems. It is worth noting that area defined by a US zip code or a UK postcode has been defined for efficient mail delivery and has only been adapted for spatial referencing purposes.

6. Spatial Data Modelling

A fundamental difference between spatial data storage structures used by different systems is the use of vector or raster data format. Digital spatial data is usually stored in one of these two formats. Vector format data uses the three basic spatial entities – points, lines and polygons (areas) – to describe topographical features. Points are stored as numbers representing their coordinate values and lines and polygons as sequences or strings (arcs, links) of coordinate points, Figures 1.3 and 1.4 display vector data. There are two main methods by which coordinates can be specified. With the absolute mode, the X,Y coordinate pair (with respect to the system origin, or sometimes a false origin) is stored. In the relative or incremental mode, only the first point on a vector is defined in absolute terms; all subsequent points are then described by an ordered sequence of relative coordinates. The relative coordinates may be either rectangular, i.e. dX, dY, or polar, i.e. angle, distance [Parker, 1990].

The raster format or grid representation of topographic data uses an infinitely repeatable set of polygons (grid cells). These regular polygons, most often rectangles, are held in the form of a mesh. Each grid cell is coded, usually numerically, to inform what type of feature that cell represents, e.g. vegetation type, height or building. In its simplest form a raster grid cell may be coded by Boolean values, i.e. true or false, Figure 1.7 (left) is such a raster grid. The term used in geometry for this is a regular tessellating. Hence, data models using this format are often referred to as tessellating models. There are various systems used for tesseral addressing, a general method is discussed by Bell *et al.* [1983].

Although object-orientation concepts have been widely adopted in computer software engineering, and commercial object-orientated databases are now available, almost all commercial GIS utilize the relational database model. This OODB approach to data modelling has appeared in the general computing arena and is now being applied to geographical models to increase their analytical effectiveness, especially where large volumes of data are involved [Gahegan, 1988; Longley *et al.*, 2001]. The creation of a data model for OOGIS is still mostly uncharted territory. Usual data modelling techniques such as entity-relationship quickly show their limitations when exposed to the needs of spatial data management. Extensions to the entity-relationship approach have been proposed, and there are also a number of techniques based on semantic data models [Gray

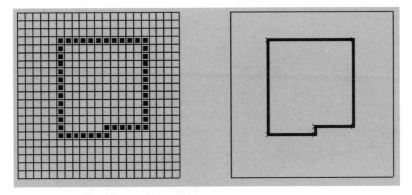

Figure 1.7   *Area feature – raster and vector forms*

et al., 1992]. The Topologically Integrated Geographic Resource Information System (TIGRIS) from Intergraph and now defunct, was based on an object-orientated data management system [Herring, 1987; Martin, 2000]. Smallworld GIS, now part of GE Network Solutions, provides an 'interactive object orientated development and customisation environment'. APIC's database engine is claimed to be capable of relating objects in the way a relational DBMS would [Davis *et al.*, 1994].

There are, of course, other data models widely accepted and defined for handling particular types of spatial features. A topological feature model maintains a record of the topology of features, i.e. which features are connected to one another, which are adjacent to one another and which features are contained inside another feature. Spatial models are also used to represent three-dimensional (3D) objects and surfaces, triangulated irregular networks (TIN), a regular grid of height points or a regular tessellation. A complex spatial model for 3D objects, using the node-relation structure to represent the topological relationships between 3D spatial objects, is given by Lee and Kwan [2000]. A comprehensive review of multidimensional GIS and related research areas is given by Raper [2000].

7.   Spatial Data Visualization

Visualization is the use of visual methods to facilitate analysis, understanding, and communication of models, concepts, and data in science and engineering. Visualization is the practice of mapping data to visual form, for exploration and analysis, for presentation. A formal definition is provided by the National Science Foundation (NSF), USA:

Figure 1.8   University of Glamorgan visualization

> Visualisation is a method of computing. It transforms the symbolic into the geometric, enabling researchers to observe their simulations and computations. Visualisation offers a method for seeing the unseen. It enriches the process of scientific discovery and fosters profound and unexpected insights.
>
> [NSF, 1987]

The visualization of geographical information and related attribute data is a fundamental role of GIS. The most common visualization tool is the VDU of a computer. More often than not, the display will be a two-dimensional (2D) digital map/diagram/surface, depicting topography, surface height, climatological variable or any other scientific, sociological or spatial statistical data. Figure 1.8 displays building footprints, extracted from Ordnance Survey Mastermap data, extruded in height, and then superimposed on aerial photography overlaid on a DTM of the area.

Modern GIS will also provide three-dimensional (3D) visualization, albeit displayed on a 2D screen, of 3D spatial data. Figure 1.9 displays a simple 3D model of buildings superimposed on a flat surface, visualized in a CAD system with shadows. Visualization of 3D models can be very realistic, without the use of actual photographs, Figure 1.10 is an example of what is possible.

8.   GIS and the Internet

The rapid growth and use of the Internet have had a significant impact on the use of GIS. The easy accessibility of data and information across the World Wide Web (WWW) has opened up a whole new era in information processing, including geographical information processing. Web-based GIS proliferates across a range of platforms using data from many disparate sources [Lake *et al.*, 2004].

Figure 1.9 *A simple 3D model*

Figure 1.10 *A realistic and futuristic 3D model of Bath*

Sources of data, previously inaccessible, are now on-line 24/7. Using data from a wide variety of independent sources does have its problems: data structure, format, compatibility, currency, accuracy, etc. A number of organizations have developed to address these issues. The World Wide Web Consortium (W3C) develops interoperable technologies (specifications, guidelines, software, and tools). W3C has developed the Geography Mark-Up Language (GML), used to describe geographic objects in a manner that can readily be shared on the Internet. In particular, GML builds on the eXtensible Mark-up Language (XML) [Lake *et al.*, 2004]. The Open GIS Consortium, Inc. (OGC) is a membership organization dedicated to the development of open system approaches to geoprocessing. 'OGC is a non-profit, international, voluntary consensus standards organization that is leading the development of standards for geospatial and location based services' [OGC, 2004].

9.   The Application of GIS

The use and analysis of geographical information are not new requirements of the modern technological age. Mankind has made use of geographical information throughout history. Military strategists, navigators, farmers and cartographers of the earliest civilizations are known to have collected and used geographical data [Burrough, 1986]. The rapid development of the industrial world, together with a greater use and understanding of the Earth's natural resources, has greatly increased the need for geographical information and its management. The variety of uses for geographical information is extensive. A few specific examples are:

- planning and management of public services;
- defense and security systems;
- land use and resource management;
- environmental monitoring;
- epidemiology;
- utility network management;
- transport network management;
- property development and investment;
- marketing and business location;
- civil engineering;
- mineral exploitation;
- vehicle navigation and tracking;
- mobile phone location-based services (LBS);
- teaching and education.

Detailed descriptions and explanations of GIS applications may be found in many GIS books, including general and broad applications in Longley *et al.* [2001], socio-economic applications in Martin [1996], archaeological applications in Wescott and Brandon [1999] and a breadth of real and experimental applications, from 1994 to the present day, in the Innovations in GIS series of books, one per year, based on contributions to the annual GIS Research UK conference, for example, book nine in the series entitled *Socio-economic Applications in Geographical Information Science* [Kidner *et al.*, 2002].

9.1. Example GIS Applications

Management of Archaeological Records. This application makes use of GIS technologies for the conservation and management of archaeological remains in the city of Newcastle upon Tyne, UK, during the period 1992 to 2000. It focuses on the use of modern technology for the capture, manipulation and visualization of archaeological remains, Figure 1.11 (Plate 3) displays the user interface to this archaeological GIS.

As part of the planning system, the Tyne and Wear archeologist requires access to a range of archaeological information from a variety of different sources. The quality of the advice provided by the archaeologists of the impact of any proposed development on the archaeological resource of the county is directly dependent on the efficiency with which information can be organized and retrieved [Taylor *et al.*, 1998]. Archaeological information exists in a variety of

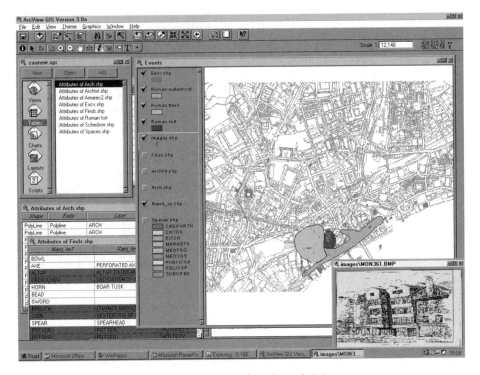

Figure 1.11 Archaeological GIS

forms obtained from many sources, ranging from modern, very complex scientific excavations to accounts of artifacts seen in the past, related to historians of dubious credibility. Traditional methods involved the use of paper records, prodigious feats of memory, supplemented by many trips to the library, with all the delay in time and expenditure of resources involved.

The objective in creating a GIS-based record was to place verified archaeological evidence within a series of databases which were directly linked to modern maps to allow spatial analysis of data and digital mapping of text-based information. Once the information is in a GIS, it becomes available for higher levels of usage. These include the construction of a variety of terrain models, virtual reality (VR), and the dissemination of data through electronic media (intranet through the local authority, WWW to the external world) and predictive modelling of ancient urban landscapes. VR provides the method of re-occupying long disappeared buildings, known to us through historical paintings and photographs.

The initial implementation of Newcastle City's Archaeological Urban Record GIS used the software known as Surveyors Land Information Management Package (SLIMPAC), and more recently ESRI's ArcView GIS, [Heslop and Taylor, 1998].

Archaeological sites are recorded as vector format graphical features stored as points, lines or closed polygons. Associated textual attribute information is held in various tables linked and joined using a relational database structure (RDBS). Additional data such as digital imagery, more recently virtual reality representations derived from antiquated photography, drawings and maps are also accessible directly from within the system. Ordnance Survey (OS) large-scale vector and raster map tiles, at various scales, are used for background mapping. Figure 1.12 shows a map created by assigning an index of archaeological potential to different areas of the city, according to how much has been destroyed, a value of 3 indicating the maximum likelihood of archaeological preservation.

A particular problem encountered during the data capture exercise was how to capture archaeological graphical detail from existing County Series mapping, not recorded in modern OS National Grid coordinates. This was solved by the use of coordinate transformation and rubber sheeting tools available in GIS. Features (usually churches) with known coordinates in both the County series historical maps and modern OS Land Line maps were identified.

The use of GIS technology for the accurate recording of position of archaeological sites from County series maps, estate plans, other maps, etc. and the capture of the data onto present-day Ordnance Survey National Grid map sheets has developed steadily over the past five years and continues to take advantage of the changes and innovations in the technology.

GIS and the Utilities. GIS utilization in a utility company (gas, electricity, telecoms and water) may be considered to be the epitome of its application. GIS, as its basic feature, provides the users with the precise geographical location of an

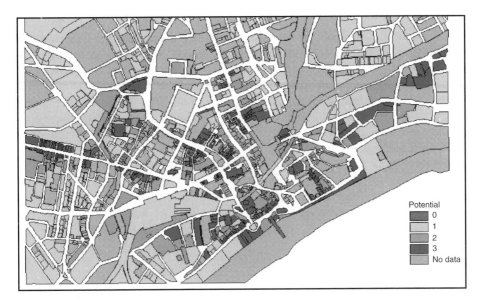

Figure 1.12   *Archaeological potential*

object. This feature of GIS is very useful to the utilities. The objects for utilities are their assets in the form of equipment and supply infrastructure that are installed in the field. Therefore, any utility would greatly benefit if it could locate its assets precisely and also get as much information about them as possible by using GIS technology. To yield this important benefit, GIS would normally require that the utility should pinpoint its assets correctly on the accurate and reliable geographical maps of its service area. To process the information associated with these assets correctly, and prepare the required reports, it is necessary that the utility includes as much information as possible, in the form of attributes of the objects that are placed on the maps. The most important need of every distribution utility is that it should be able to conduct reliable network analysis studies that will compute the system's technical losses correctly. It is also important for the utilities that they are able to work out practicable and affordable network improvement schemes, based on the results of these studies. These schemes should preferably be such that when implemented in the field, they will not only reduce the current technical losses in the system but will also lead to overall improvement of the present performance of the supply networks [Piplapure, 2004].

10.   Conclusion

GIS can be as simple as relatively simple electronic map browsers or as complex as complex spatial data processing and analysis, Spatial Decision Support

Systems (SDSS), and a great deal in between. This chapter has merely touched on the subject of GIS basics. It has introduced the main concepts and given many references to literature sources that provide much more detailed descriptions and explanations of the GI topics mentioned. It has provided an explanation and definition of GIS and geographical data, explained the basic elements of spatial and aspatial data, spatial referencing, features, and attributes. Furthermore, this chapter has introduced spatial data modelling and visualization. Two typical GIS applications were briefly discussed, one a land information system and the other a network analysis system.

2

GPS: An Introduction

The aim of this chapter is to introduce the principles of GPS theory, and to provide a background for later chapters. With that in mind, some of the theoretical treatment has been simplified to provide a starting point for a mathematically literate user of GPS who wants to understand how GPS works, and to get a basic grasp of GPS theory and terminology.

After this book is published, other Global Navigation Satellite Systems (GNSS) will become fully operational, such as the Russian GLONASS system or the planned European Galileo system. Although the technical details of other GNSS systems might differ, there will certainly be much overlap in the fundamental principles presented here.

The fundamental principle to keep in mind is that all GNSS systems are timing systems. By use of precise timing information on radio waves transmitted from the GNSS satellite, the user's receiver can measure the range to each satellite in view, and hence calculate its position. Positions can be calculated at every measurement epoch, which may be once per second when applied to car navigation. Kinematic parameters such as velocity and acceleration are secondary, in that they can then be calculated from the measured time series of positions.

1. GPS Description

1.1. The Basic Idea

GPS positioning is based on trilateration, which is the method of determining position by measuring distances to points at known coordinates. At a minimum, trilateration requires three ranges to three known points. GPS point positioning, on the other hand, requires four 'pseudoranges' to four satellites. This raises two questions: (1) 'What are pseudoranges?', and (2) 'How do we know the position of the satellites?' Without going into too much detail at this point, we shall address the second question first.

Intelligent Positioning: GIS-GPS Unification G. Taylor and G. Blewitt
© 2006 John Wiley & Sons, Ltd

How Do We Know the Position of the Satellites? A signal is transmitted from each satellite towards the Earth. This signal is encoded with the 'Navigation Message', which can be read by the user's GPS receiver. The Navigation Message includes orbit parameters (often called the 'Broadcast Ephemeris'), from which the receiver can compute satellite coordinates (X,Y,Z). These are Cartesian coordinates in a geocentric system, known as WGS-84, which has its origin at the Earth centre of mass, Z axis pointing towards the North Pole, X pointing towards the Prime Meridian (which crosses Greenwich), and Y at right angles to X and Z to form a right-handed orthogonal coordinate system. The algorithm which transforms the orbit parameters into WGS-84 satellite coordinates at any specified time is called the 'Ephemeris Algorithm', which is defined in GPS textbooks [e.g., Leick, 1990]. We shall discuss the Navigation Message in more detail later on; now, we move on to 'pseudoranges'.

What Are Pseudoranges? The time at which the signal is transmitted from the satellite is encoded on the signal, using the time according to an atomic clock on board the satellite. The time of signal reception is recorded by a receiver using its own low precision clock. The satellites have highly precise atomic clocks. A receiver measures the difference in these times:

$$\text{pseudorange} = (\text{time difference}) \times (\text{speed of light})$$

Note that a pseudorange is almost like a range, except that it includes clock and other errors because the receiver clocks are far from perfect. How do we correct for clock errors?

How Do We Correct for Clock Errors? The clock error for each satellite is given in the corresponding Navigation Message, in the form of a number of polynomial coefficients. The unknown receiver clock error can be estimated by the user along with unknown station coordinates. Hence, there are four unknowns, requiring a minimum of four pseudorange measurements. These four unknowns are the three-dimensional position in space and the receiver clock error.

1.2.   The GPS Segments

There are four GPS segments:

1 The Space Segment, which includes the constellation of GPS satellites, which transmit the signals to the user.
2 The Control Segment, which is responsible for the monitoring and operation of the Space Segment.
3 The User Segment, which includes user hardware and processing software for positioning, navigation, and timing applications.
4 The Ground Segment, which includes civilian tracking networks that provide the User Segment with reference control, precise ephemerides, and real-time services (DGPS) which mitigate the effects of 'selective availability' (a topic discussed later).

Before getting into the details of the GPS signal, observation models, and position computations, we first provide more information on the Space Segment and the Control Segment.

Orbit Design. The satellite constellation is designed to have at least four satellites in view anywhere, anytime, to a user on the ground. For this purpose, there are nominally 24 GPS satellites distributed in six orbital planes. So that we may discuss the orbit design and the implications of that design, we must digress briefly to explain the geometry of the GPS constellation.

According to Kepler's laws of orbital motion, each orbit takes the approximate shape of an ellipse, with the Earth's centre of mass at the focus of the ellipse. For a GPS orbit, the eccentricity of the ellipse is so small (0.02) that it is almost circular. The semi-major axis (largest radius) of the ellipse is approximately 26,600 km, or approximately four Earth radii.

The six orbital planes rise over the equator at an inclination angle of 55°. The point at which they rise from the Southern to Northern Hemisphere across the equator is called the 'Right Ascension of the ascending node'. Since the orbital planes are evenly distributed, the angle between the six ascending nodes is 60°.

Each orbital plane nominally contains four satellites, which are generally not spaced evenly around the ellipse. Therefore, the angle of the satellite within its own orbital plane, the 'true anomaly', is only approximately spaced by 90°. The true anomaly is measured from the point of closest approach to the Earth (the perigee). (We note here that there are other types of 'anomaly' in GPS terminology, which are angles that are useful for calculating the satellite coordinates within its orbital plane.) Note that instead of specifying the satellite's anomaly at every relevant time, we could equivalently specify the time that the satellite had passed the perigee, and then compute the satellite's future position based on the known laws of motion of the satellite around an ellipse.

Finally, the argument of perigee is the angle between the equator and perigee. Since the orbit is nearly circular, this orbital parameter is not well defined, and alternative parameterization schemes are often used.

Taken together (the eccentricity, semi-major axis, inclination, Right Ascension of the ascending node, the time of perigee passing, and the argument of perigee), these six parameters define the satellite orbit. These parameters are known as Keplerian elements. Given the Keplerian elements and the current time, it is possible to calculate the coordinates of the satellite.

GPS satellites do not move in perfect ellipses, so additional parameters are necessary. Nevertheless, GPS does use Kepler's laws to its advantage, and the orbits are described in parameters that are Keplerian in appearance. Additional parameters must be added to account for non-Keplerian behaviour. Even this set of parameters has to be updated by the Control Segment every hour for them to remain sufficiently valid.

Orbit Design Consequences. Several consequences of the orbit design can be deduced from the above orbital parameters, and Kepler's laws of motion. First of all, the satellite speed can be easily calculated to be approximately 4 km/s relative to Earth's centre. All the GPS satellites orbits are prograde, which means the satellites move in the direction of the Earth's rotation. Therefore, the relative motion between the satellite and a user on the ground must be less than 4 km/s. Typical values around 1 km/s can be expected for the relative speed along the line of sight (range rate).

The second consequence is the phenomena of 'repeating ground tracks' every day. It is straightforward to calculate the time it takes for the satellite to complete one orbital revolution. The orbital period is approximately T = 11 hr 58 min. Therefore, a GPS satellite completes two revolutions in 23 hr 56 min. This is intentional, since it equals the sidereal day, which is the time it takes for the Earth to rotate 360°. (Note that the solar day of 24 hr is not 360°, because during the day, the position of the Sun in the sky has changed by 1/365.25 of a day, or 4 min, due to the Earth's orbit around the Sun.)

Therefore, every day (minus 4 minutes), the satellite appears over the same geographical location on the Earth's surface. The 'ground track' is the locus of points on the Earth's surface that is traced out by a line connecting the satellite to the centre of the Earth. The ground track is said to repeat. From the user's point of view, the same satellite appears in the same direction in the sky every day minus 4 minutes. Likewise, the 'sky tracks' repeat. In general, we can say that the entire satellite geometry repeats every sidereal day (from the point of view of a ground user).

As a corollary, any errors correlated with satellite geometry will repeat from one day to the next. An example of an error tied to satellite geometry is 'multipath', which is due to the antenna also sensing signals from the satellite that reflect and refract from nearby objects. In fact, it can be verified that, because of multipath, observation residuals do have a pattern that repeats every sidereal day. As a consequence, such errors will not significantly affect the precision, or repeatability, of coordinates estimated each day. However, the accuracy can be significantly worse than the apparent precision for this reason.

Another consequence of this is that the same subset of the 24 satellites will be observed every day by someone at a fixed geographical location. Generally, not all 24 satellites will be seen by a user at a fixed location. This is one reason why there needs to be a global distribution of receivers around the globe to be sure that every satellite is tracked sufficiently well.

We now turn our attention to the consequences of the inclination angle of 55°. Note that a satellite with an inclination angle of 90° would orbit directly over the poles. Any other inclination angle would result in the satellite never passing over the poles. From the user's point of view, the satellite's sky track would never cross over the position of the celestial pole in the sky. In fact, there would be a 'hole' in the sky around the celestial pole where the satellite could never pass. For a satellite constellation with an inclination angle of 55°, there would therefore be a circle of radius at least 35° around the celestial pole, through which the sky tracks would never cross. This has a significant effect on the satellite geometry as viewed from different latitudes. An observer at the pole would never see a GPS satellite rise above 55° elevation. Most of the satellites would hover close to the horizon. Therefore vertical positioning is slightly degraded near the poles. An observer at the equator would see some of the satellites passing overhead, but they would tend to deviate from away from points on the horizon directly to the north and south. Due to a combination of Earth rotation, and the fact that the GPS satellites are moving faster than the Earth rotates, the satellites actually appear to move approximately north–south or south–north to an oberver at the equator, with very little east–west motion. Therefore, the closer the observer is to the equator, the better determined becomes the north component of relative position as compared to the east component. An observer at mid-latitudes in the Northern Hemisphere would see satellites anywhere in the sky to the south, but there would be a large void towards the north. This has consequences for site selection, where a good view is desirable to the south, and the view to the north is less critical. For example, one might want to select a site in the Northern Hemisphere that is on a south-facing slope (and vice versa for an observer in the Southern Hemisphere).

Satellite Hardware. There are nominally 24 GPS satellites, but this number can vary within a few satellites at any given time, due to old satellites being decommissioned, and new satellites being launched to replace them. All the prototype satellites, known as Block I, have been decommissioned. Between 1989 and 1994, 24 Block II (1989–1994) were placed in orbit. From 1995 onwards, these have started to be replaced by a new design known as Block IIR. The nominal specifications of the GPS satellites are as follows:

1 Life goal: 7.5 years.
2 Mass: ~1 tonne (Block IIR: ~2 tonnes).
3 Size: 5 metres.
4 Power: solar panels $7.5\,m^2$ + Ni-Cd batteries.
5 Atomic clocks: 2 rubidium and 2 cesium.

The orientation of the satellites is always changing, so that the solar panels face the sun, and the antennas face the centre of the Earth. Signals are transmitted and received by the satellite using microwaves. Signals are transmitted to the User Segment at frequencies L1 = 1575.42 MHz, and L2 = 1227.60 MHz. We discuss the signals in further detail later on. Signals are received from the Control Segment at frequency 1783.74 Mhz. The flow of information is as follows: the satellites transmit L1 and L2 signals to the user, which are encoded with information on their clock times and their positions. The Control Segment then tracks these signals using receivers at special monitoring stations. This information is used to improve the satellite positions and predict where the satellites will be in the near future. This orbit information is then uplinked at 1783.74 Mhz to the GPS satellites, which in turn transmit this new information down to the users, and so on. The orbit information on board the satellite is updated every hour.

The Control Segment. The Control Segment, run by the US Air Force, is responsible for operating GPS. The main Control Centre is at Falcon Air Force Base, Colorado Springs, USA. Several ground stations monitor the satellites' L1 and

L2 signals, and assess the 'health' of the satellites. As outlined previously, the Control Segment then uses these signals to estimate and predict the satellite orbits and clock errors, and this information is uploaded to the satellites. In addition, the Control Segment can control the satellites; for example, the satellites can be manoeuvred into a different orbit when necessary. This might be done to optimize satellite geometry when a new satellite is launched, or when an old satellite fails. It is also done to keep the satellites to within a certain tolerance of their nominal orbital parameters (e.g. the semi-major axis may need adjustment from time to time). As another example, the Control Segment might switch between the several on-board clocks available, should the current clock appear to be malfunctioning.

1.3.   The GPS Signals

We now briefly summarize the characteristics of the GPS signals, the types of information that is digitally encoded on the signals, and how the US Department of Defense (DOD) implements denial of accuracy to civilian users. Further details on how the codes are constructed will be presented in Section 2.

Signal Description. The signals from a GPS satellite are basically driven by an atomic clocks (usually cesium, which has the best long-term stability). The fundamental frequency is 10.23 MHz. Two carrier signals, which can be thought of as sine waves, are created from this signal by multiplying the frequency by 154 for the L1 channel (frequency = 1575.42 MHz; wavelength = 19.0 cm), and 120 for the L2 channel (frequency = 1227.60 MHz; wavelength = 24.4 cm). The reason for the second signal is for self-calibration of the delay of the signal in the Earth's ionosphere.

Information is encoded in the form of binary bits on the carrier signals by a process known as *phase modulation*. (This is to be compared with signals from radio stations, which are typically encoded using either frequency modulation, FM, or amplitude modulation, AM.) The binary digits 0 and 1 are actually represented by multiplying the electrical signals by either +1 or −1, which is equivalent to leaving the signal unchanged, or flipping the phase of the signal by 180°. We come back later to the meaning of phase and the generation of the binary code.

There are three types of code on the carrier signals:

1 The C/A (course acquisition) code.
2 The P (precise) code.
3 The Navigation Message.

The C/A ('course acquisition') code can be found on the L1 channel. As will be described later, this is a code sequence which repeats every 1 ms. It is a pseudo-random code, which appears to be random, but is in fact generated by a known algorithm. The carrier can transmit the C/A code at 1.023 Mbps (million bits per second). The 'chip length', or physical distance between binary transitions (between digits +1 and −1), is 293 metres. The basic information that the C/A code contains is the time according to the satellite clock when the signal was transmitted (with an ambiguity of 1 ms, which is easily resolved, since this corresponds to 293 km). Each satellite has a different C/A code, so that they can be uniquely identified.

The P ('precise') code is identical on both the L1 and L2 channel. Whereas C/A is a coarser code appropriate for initially locking onto the signal, the P code is better for more precise positioning. The P code repeats every 267 days. In practice, this code is divided into 7-day segments; each weekly segment is designated a 'PRN' number, and is designated to one of the GPS satellites. The carrier can transmit the P code at 10.23 Mbps, with a chip length of 29.3 metres. Again, the basic information is the satellite clock time or transmission, which is identical to the C/A information, except that it has ten times the resolution. Unlike the C/A code, the P code can be encrypted by a process known as 'anti-spoofing', or 'A/S' (see below).

The Navigation Message can be found on the L1 channel, being transmitted at a very slow rate of 50 bps. It is a 1500 bit sequence, and therefore takes 30 seconds to transmit. The Navigation Message includes information on the Broadcast Ephemeris (satellite orbital parameters), satellite clock corrections, almanac data (a crude ephemeris for all satellites), ionosphere information, and satellite health status.

Denial of Accuracy.   The U.S. Department of Defense can implement two types of denial of accuracy to civilian users: Selective Availability (S/A) and Anti-Spoofing (A/S). S/A can be thought of as intentional errors imposed on the GPS signal. A/S can be thought of as encryption of the P code.

There are two types of S/A: 'epsilon' and 'dither'. Under conditions of S/A, the user should be able to count on the position error not being any worse than 100 metres. Most of the time, the induced position errors do not exceed 50 metres.

'Epsilon' is implemented by including errors in the satellite orbit encoded in the Navigation Message. Apparently, this is an option not used, according to daily comparisons made between the real-time broadcast orbits, and precise orbits generated after the fact, by the International GNSS Service for Geodynamics (IGS). For precise geodetic work, precise orbits are recommended in any case, and therefore epsilon would have minimal impact on precise users. It would, however, directly impact single receiver, low-precision users. Even then, the effects can be mitigated to some extent by using technology known as 'differential GPS', where errors in the GPS signal are computed at a reference station at known coordinates, and are transmitted to the user who has appropriate radio receiving equipment.

'Dither' is intentional rapid variation in the satellite clock frequency (10.23 MHz). Dither, therefore, looks exactly like a satellite clock error, and therefore maps directly into pseudorange errors. As is the case for epsilon, dither can be mitigated using differential GPS. The high precision user is minimally effected by S/A, since relative positioning techniques effectively eliminate satellite clock error (as we shall see later). For civilian and economic reasons, dither was switched off on 2 May 2000 by executive order of President Bill Clinton. Since dither is nothing more than a satellite clock effect, any steps taken to mitigate dither will also mitigate satellite clock errors.

Anti-Spoofing (A/S) is encryption of the P code. The main purpose of A/S is to prevent 'the enemy' from imitating a GPS signal, and therefore it is unlikely to be switched off in the foreseeable future. A/S does not pose a significant problem to the precise user, since precise GPS techniques rely on measuring the phase of the carrier signal itself, rather than the pseudoranges derived from the P code. However, the pseudoranges are very useful for various algorithms, particularly in the rapid position fixes required by moving vehicles and kinematic surveys. Modern geodetic receivers can, in any case, form two precise pseudorange observables on the L1 and L2 channels, even if A/S is switched on (we briefly touch on how this is done in the next section). As a consequence of not having full access to the P code, the phase noise on measuring the L2 carrier phase can be increased from the level of 1 mm to the level of 1 cm for some types of receivers. This has negligible impact on long sessions for static positioning, but can have a noticeable effect on short sessions, or on kinematic positioning. Larger degradation in the signal can be expected at low elevations (up to 2 cm) where signal strength is at a minimum.

2.   The Pseudorange Observable

In this section, we go deeper into the description of the pseudorange observable, and give some details on how the codes are generated. We develop a model of the pseudorange observation, and then use this model to derive a least-squares estimator for positioning. We discuss formal errors in position, and the notion of 'Dilution of Precision', which can be used to assess the effect of satellite geometry on positioning precision.

2.1.   Code Generation

It helps to understand the pseudorange measurement if we first take a look at the actual generation of the codes. The carrier signal is multiplied by a series of either +1 or −1, which are seperated by the chip length (293 m for C/A code, and 29.3 m for P code). This series of +1 and −1 multipliers can be interpreted as a stream of binary digits (0 and 1). How is this stream of binary digits decided? They are determined by an algorithm, known as a linear feedback register. To understand a linear feedback register, we must first introduce the XOR binary function.

XOR: The 'Exclusive OR' Binary Function. A binary function takes two input binary digits, and outputs one binary digit (0 or 1). More familiar binary functions might be the 'AND' and 'OR' functions. For example, the AND function gives a value of 1 if the two input digits are identical, that is (0, 0), or (1, 1). If the input digits are different, the AND function gives a value of 0. The OR function gives a value of 1 if either of the two input digits equals 1, that is (0, 1), (1, 0), or (1, 1). The XOR function gives a value of 1 if the two inputs are different, that is (1, 0) or (0, 1). If the two inputs are the same, (0, 0) or (1, 1), then the value is 0.

What is XOR(A,B)? Remember this: *Is A different to B? If so, the answer is 1*:

$$\text{If } A \neq B, \text{ then XOR(A,B)} = 1.$$
$$\text{If } A = B, \text{ then XOR(A,B)} = 0.$$

The XOR function can be represented by the 'truth table' shown in Table 2.1.

Linear Feedback Registers. Linear feedback registers are used to generate a pseudorandom number sequence. The sequence is pseudorandom, since the sequence repeats after a certain number of digits (which, as we shall see, depends on the size of the register). However, the statistical properties of the sequence are very good, in that the sequence appears to be white noise. We return to these properties later, since they are important for understanding the measurement process. For now, we look at how the register works.

Table 2.2 illustrates a simple example: the '3-stage linear feedback register'. The 'state' of the register is defined by three binary numbers (A, B, C). The state changes after a specific time interval. To start the whole process, the intial state of a feedback register is always filled with 1, that is, for the 3-stage register, the

Table 2.1 *Truth table for the XOR function*

Input A	Input B	Output XOR(A,B)
0	0	0
0	1	1
1	0	1
1	1	0

Table 2.2 *A 3-stage linear feedback register*

Cycle, N	$A_N = \text{XOR}(A_{N-1}, C_{N-1})$	$B_N = A_{N-1}$	$C_N = B_{N-1}$
1	initialize: 1	1	1
2	XOR(1,1) = 0	1	1
3	XOR(0,1) = 1	0	1
4	XOR(1,1) = 0	1	0
5	XOR(0,0) = 0	0	1
6	XOR(0,1) = 1	0	0
7	XOR(1,0) = 1	1	0
8 (= 1)	XOR(1,0) = 1	1	1
	(pattern repeats)		

initial state is $(1, 1, 1)$. The digits in this state are now shifted to the right, giving (blank, $1, 1$). The digit (1) that is pushed off the right side is the output from the register. The blank is replaced by taking the XOR of the output and first digit $(1, 1)$. The value, in this case, equals 0. The new state is therefore $(0, 1, 1)$. This process is then repeated, so that the new output is (1), and the next state is $(1, 0, 1)$. The next output is (1) and the next state is $(1, 1, 0)$. The next output is (0), and the next state is $(0, 1, 1)$, and so on.

In the above example, the outputs can be written as $(1, 1, 1, 0, \ldots)$. This stream of digits is known as the 'linear feedback register sequence'. This sequence will start to repeat after a while. It turns out that during a complete cycle, the feedback register will produce every possible combination of binary numbers, except for $(0, 0, 0)$. We can therefore easily calculate the length of the sequence before it starts to repeat again. For a 3-stage register, there are eight possible combinations of binary digits. This means that the sequence will repeat after seven cycles. The sequence length is therefore 7 bits. More generally, the sequence length is:

$$L(N) = 2^N - 1$$

where N is the size of the register (number of digits in the state). For example, a 4 state linear feedback register will have a sequence length of 15 bits.

C/A Code. The C/A code is based on the 10-stage linear feedback register sequence, for which the sequence length is $L(10) = 2^{10} - 1 = 1023$ bits. The C/A code really has a repeating sequence of 1023 bits, however, the design is slightly more complicated than presented above. The C/A code is actually a 'Gold code', which is derived by taking the XOR of the output from two linear feedback registers. Unique C/A codes can be generated for each satellite by selecting different pairs of cells from each register to define the output.

In summary, the C/A code is a unique Gold code on each satellite, which is a pseudorandom sequence of bits with a repeating sequence length of 1023. C/A bit transitions occur at 1.023 MHz. Note that the fundamental frequency in the satellite is 10.23 MHz, so this represents one transition every 10 cycles. At this rate of bit transitions, the full sequence of 1023 bits is transmitted in 1 ms. Therefore, the sequence repeats 1000 times per second. The chip length (distance between bit transitions) is 293 m. Therefore, the sequence repeats every 300 km.

P Code. The P code is also generated from a combination of two different registers, in such a way that it repeats every 266.4 days. Each 7-day section is assigned a 'PRN code'. Satellites are often identified by their PRN number; however, the user should note that any given satellite can have its PRN code changed. Therefore, PRN codes should not be used in place of Satellite Vehicle Numbers (SVN) when talking about particular satellites. (For example, if someone writes software which identifies satellites using PRN numbers, there might be a problem in orbit modelling, for example, PRN 2 is assigned to a Block II satellite now, but to a Block IIR satellite next year.) There are 38 possible PRN codes; given that there are 24 nominal satellites, some PRN codes are left unused. The PRN sequence is reset at Saturday midnight, defining the start of 'GPS week'.

GPS Signal Transmission and Reception. Let us now summarize how the GPS signal is transmitted from space, and then received on the ground. The GPS signal starts in the satellite as a voltage, which oscillates at the fundamental clock frequency of 10.23 MHz. (If S/A is on, this signal is then 'dithered' so that the frequency varies unpredictably.) This signal is then separately multiplied in frequency by the integers 154 and 120, to create the L1 and L2 carrier signals. The signals are then multiplied by +1 and −1 according the algorithms described above to generate the C/A code (on L1) and the P code (on both L1 and L2). These codes are unique to each satellite. Finally, the Navigation Message is encoded onto the signal. The signals are boosted by an amplifier, and then sent to transmitting antennas, which point towards the Earth. These antennas are little more than exposed electrical conductors which radiate the signal into space in the form of electromagnetic waves.

These electromagnetic waves pass through space and the Earth's atmosphere, at very close to the speed of light in a vacuum, until they reach the receiver's antenna. The waves create a minute signal in the antenna, in the form of an oscillating voltage. The signal is now pre-amplified at the antenna, to boost the signal strength, so that it is not overcome by noise by the time it gets to the other end of the antenna cable. The signal then enters the receiver, which then measures it using a process known as 'autocorrelation'. It is beyond the scope of this book to go into the details of receiver design, so our description will be kept at the level required to understand how the observable model can be developed.

2.2.   Autocorrelation Technique

We have described how the GPS satellites construct the GPS signals. Actually, the receiver also generates GPS-like signals internally. The receiver knows precisely what the transmitted GPS signal is supposed to look like at any given time, and it generates an electronic replica, in synchronization with the receiver's own clock. The receiver then compares the replica signal with the actual signal. Since the GPS signal was actually created in the satellite some time previously (about 0.07 seconds ago, due to the speed of light), the receiver's replica signal must be delayed to match up the incoming signal with the replica signal. This time delay is actually what the receiver is fundamentally measuring. Clearly, this represents the time taken for the signal to pass from the satellite to the receiver, but it also includes any error in the satellite clock, and any error in the receiver clock. One can see that the time delay is therefore related to the range to the satellite. We return to this model later, and now turn our attention to how the receiver matches the two signals.

The time difference is computed by *autocorrelation*. The first bit from signal one is multiplied by the first bit of signal two. For example, if the first bits from the two signals both have values −1, then the result is $(-1) \times (-1) = +1$. Similarly, if both bits have values +1, then the result is +1. On the other hand, if the two bits disagree, the result is $(+1) \times (-1) = -1$. This process is repeated for the second pair of bits, and so on. The result can be written as a sequence of +1 (where the bits agree) and –1 (where the bits disagree). This sequence is then summed, and divided by the total number of bits in each signal. For example, if signal A can be written (+1, –1, –1, +1, –1), and signal B can be written (+1, +1, –1, –1, +1), then multiplication gives (+1, –1, +1, –1, –1); the sum of which gives –1; then dividing by the number of bits (5) gives –0.2. Note that if the two signals matched perfectly, the result would be +1. If the two signals were completely random, we should expect a result close to zero.

This is why the GPS signals are designed to look random. When the two signals are not properly matched in time, the result of autocorrelation gives an answer close to zero; if the signals are matched in time, the result is close to +1 (but not exactly, since a real signal also has noise, so some bits are incorrect). One can see that the larger the number of bits that are compared, the better the resolution. This is because the random bits will average to zero better, the more bits we compare.

The Gold codes have the property that the autocorrelation is constant until we get to within one chip of the correct answer. Within that window of ±1 chip, the autocorrelation function looks like an equilateral triangle, with a value of 1 at its peak (assuming no noise). We can therefore use the known triangular shape as a model to help us find the time displacement that maximizes the autocorrelation. (More sophisticated receivers account for the fact that multipath distorts the shape of this triangle, and can thus reduce the effect of multipath.)

Now that we have found the peak autocorrelation, the inferred time displacement between the two signals is multiplied by the speed of light. This observation is called the *pseudorange*. The pseudorange measurement is shown in Figure 2.1.

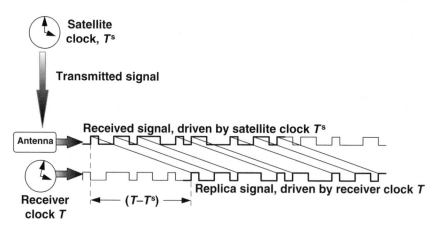

Figure 2.1   *How the GPS pseudorange observation is related to the satellite and receiver clocks*

2.3.   Pseudorange Observation Equations

GPS receivers record data at regular, user-specified intervals (say, every 30 seconds). It is the reading of the receiver clock time T that is used to say exactly when the measurement is sampled. Therefore, the value of T at a measurement epoch is known exactly, and is written to the data file along with the observation. (What is not known, is the true time of measurement.) The actual observation to satellites can be writted:

$$P^S = (T - T^S)c$$

where T is the known reading of the receiver clock when signal is received, T^S is the reading of the satellite clock when the signal was transmitted, and c is the speed of light (in a vacuum) = 299792458 m/s.

The modelled observation can be developed by setting the clock time T equal to the true reception time t plus a clock bias τ, for both the receiver and satellite clocks:

$$T = t + \tau$$
$$T^S = t^S + \tau^S$$

Substitution gives the pseudorange as a function of the true time the signal was received:

$$P^S(t) = ((t + \tau) - (t^S + \tau^S))c$$
$$= (t - t^S)c + c\tau - c\tau^S$$
$$= \rho^S(t, t^S) + c\tau - c\tau^S$$

where $\rho^S(t, t^S)$ is the range from receiver (at reception time) to the satellite (at transmit time). This model is simplified; for example, it assumes the speed of light in the atmosphere is c, and it ignores the theory of relativity; but this simplified model is useful to gain insight into the principles of GPS. From Pythagoras' theorem, we can write:

$$\rho^S(t, t^S) = \sqrt{\left(x^S(t^S) - x(t)\right)^2 + \left(y^S(t^S) - y(t)\right)^2 + \left(z^S(t^S) - z(t)\right)^2}$$

The Navigation Message allows us to compute the satellite position (x^S, y^S, z^S) and the satellite clock bias τ^S. Therefore we are left with four unknowns, the receiver position (x, y, z) and the receiver clock bias τ.

We note here one complication: that the satellite position must be calculated at transmission time, t^S. This is important, because the satellite range can change as much as 60 metres from the time the signal was transmitted, to the time the signal was received, approximately 0.07 seconds later. If the reception time were used instead, the error in computed range could be tens of metres. Starting with the reception time, t, the transmit time can be computed by an iterative algorithm known as 'the light time equation', which can be written as follows:

$$t^S(0) = t = (T - \tau)$$

$$t^S(1) = t - \frac{\rho^S(t, t^S(0))}{c}$$

$$t^S(2) = t - \frac{\rho^S(t, t^S(1))}{c}$$

$$\vdots$$

where the satellite position (and hence the range $\rho^S(t, t^S)$) is calculated at each step using the Keplerian-type elements from the Navigation Message, and the algorithm is stopped once the computed range converges (i.e., don't change by more than a negligible amount). Although more rapidly converging methods have been implemented, the above method is probably the easiest to understand.

Note that the above algorithm starts with the true reception time, which requires the receiver clock bias. We usually don't know in advance what the bias is, but for most receivers it never is larger than a few milliseconds (beyond which, the receiver will reset its clock). If we assume it is zero in the above computation, the error produced is a few metres. We can therefore safely ignore this effect for now, and return to it later when we discuss the more precise carrier phase observable.

We now look at our system of simplified observation equations from four satellites in view of the receiver. Using the above notation, we can write the pseudoranges to each satellite as:

$$P^1 = ((x^1 - x)^2 + (y^1 - y)^2 + (z^1 - z)^2)^{1/2} + c\tau - c\tau^1$$
$$P^2 = ((x^2 - x)^2 + (y^2 - y)^2 + (z^2 - z)^2)^{1/2} + c\tau - c\tau^2$$
$$P^3 = ((x^3 - x)^2 + (y^3 - y)^2 + (z^3 - z)^2)^{1/2} + c\tau - c\tau^3$$
$$P^4 = ((x^4 - x)^2 + (y^4 - y)^2 + (z^4 - z)^2)^{1/2} + c\tau - c\tau^4$$

(Note that in this and subsequent equations, the superscripts next to the satellite coordinates are meant to identify the satellite, and should not be confused with exponents.) In the following section, we proceed to solve this system of equations for the four unknowns, (x, y, z, τ) using familiar least squares methods. Although this is not strictly necessary for four unknowns with four parameters, it does generalize the solution to the case where we have m $\geq$ 4 satellites in view.

3.   Point Positioning Using Pseudorange

3.1.   Least Squares Estimation

Linearized Model. We solve the point positioning problem first by linearizing the pseudorange observation equations, and then using the familiar methods of least squares analysis. For completeness, we summarize the linearization procedure and the development of the least squares method specifically for the GPS point positioning problem. First, we assume we can write the actual observation to be the sum of a modelled observation, plus an error term:

$$P_{\text{observed}} = P_{\text{model}} + \text{noise}$$
$$= P(x, y, z, \tau) + v$$

Next, we apply Taylor's theorem, where we expand about the model computed using provisional parameter values (x_0, y_0, z_0, τ_0), and ignore second and higher order terms.

$$P(x, y, z, \tau) \cong P(x_0, y_0, z_0, \tau_0) + (x - x_0)\frac{\partial P}{\partial x} + (y - y_0)\frac{\partial P}{\partial y} + (z - z_0)\frac{\partial P}{\partial z} + (\tau - \tau_0)\frac{\partial P}{\partial \tau}$$

$$= P_{\text{computed}} + \frac{\partial P}{\partial x}\Delta x + \frac{\partial P}{\partial y}\Delta y + \frac{\partial P}{\partial z}\Delta z + \frac{\partial P}{\partial \tau}\Delta \tau$$

Note that the partial derivatives in the above expression are also computed using provisional values (x_0, y_0, z_0, τ_0). The residual observation is defined to be the difference between the actual observation and the observation computed using the provisional parameter values:

$$\Delta P \equiv P_{\text{observed}} - P_{\text{computed}}$$

$$= \frac{\partial P}{\partial x}\Delta x + \frac{\partial P}{\partial y}\Delta y + \frac{\partial P}{\partial z}\Delta z + \frac{\partial P}{\partial \tau}\Delta \tau + v$$

This can be written in matrix form:

$$\Delta P = \begin{pmatrix} \dfrac{\partial P}{\partial x} & \dfrac{\partial P}{\partial y} & \dfrac{\partial P}{\partial z} & \dfrac{\partial P}{\partial \tau} \end{pmatrix} \begin{pmatrix} \Delta x \\ \Delta y \\ \Delta z \\ \Delta \tau \end{pmatrix} + v$$

We get such an equation for each satellite in view. In general, for m satellites, we can write this system of m equations in matrix form:

$$\begin{pmatrix} \Delta P^1 \\ \Delta P^2 \\ \Delta P^3 \\ \vdots \\ \Delta P^m \end{pmatrix} = \begin{pmatrix} \dfrac{\partial P^1}{\partial x} & \dfrac{\partial P^1}{\partial y} & \dfrac{\partial P^1}{\partial z} & \dfrac{\partial P^1}{\partial \tau} \\ \dfrac{\partial P^2}{\partial x} & \dfrac{\partial P^2}{\partial y} & \dfrac{\partial P^2}{\partial z} & \dfrac{\partial P^2}{\partial \tau} \\ \dfrac{\partial P^3}{\partial x} & \dfrac{\partial P^3}{\partial y} & \dfrac{\partial P^3}{\partial z} & \dfrac{\partial P^3}{\partial \tau} \\ \vdots & \vdots & \vdots & \vdots \\ \dfrac{\partial P^m}{\partial x} & \dfrac{\partial P^m}{\partial y} & \dfrac{\partial P^m}{\partial z} & \dfrac{\partial P^m}{\partial \tau} \end{pmatrix} \begin{pmatrix} \Delta x \\ \Delta y \\ \Delta z \\ \Delta \tau \end{pmatrix} + \begin{pmatrix} v^1 \\ v^2 \\ v^3 \\ \vdots \end{pmatrix}$$

The equation is often written using matrix symbols as:

$$\mathbf{b} = \mathbf{Ax} + \mathbf{v}$$

which expresses a linear relationship between the residual observations **b** (i.e., observed minus computed observations) and the unknown correction to the parameters **x**. The column matrix **v** contains all the noise terms, which are also unknown at this point. We call the above matrix equation the 'linearized observation equations'.

The Design Matrix. The linear coefficients, contained in the 'design matrix' $\mathbf{A}$, are actually the partial derivatives of each observation with respect to each parameter, computed using the provisional parameter values. Note that $\mathbf{A}$ has the same number of columns as there are parameters, $n = 4$, and has the same number of rows as there are data, $m \geq 4$. We can derive the coefficients of $\mathbf{A}$ by partial differentiation of the observation equations, producing the following expression:

$$\mathbf{A} = \begin{pmatrix} \dfrac{x_0 - x^1}{\rho} & \dfrac{y_0 - y^1}{\rho} & \dfrac{z_0 - z^1}{\rho} & c \\ \dfrac{x_0 - x^2}{\rho} & \dfrac{y_0 - y^2}{\rho} & \dfrac{z_0 - z^2}{\rho} & c \\ \dfrac{x_0 - x^3}{\rho} & \dfrac{y_0 - y^3}{\rho} & \dfrac{z_0 - z^3}{\rho} & c \\ \vdots & \vdots & \vdots & \vdots \\ \dfrac{x_0 - x^m}{\rho} & \dfrac{y_0 - y^m}{\rho} & \dfrac{z_0 - z^m}{\rho} & c \end{pmatrix}$$

Note that $\mathbf{A}$ is shown to be purely a function of the direction to each of the satellites as observed from the receiver.

The Least Squares Solution. Let us consider a solution for the linearized observation equations, $\hat{\mathbf{x}}$. The 'estimated residuals' are defined as the difference between the actual observations and the new, estimated model for the observations. Using the linearized form of the observation equations, we can write the estimated residuals as:

$$\hat{\mathbf{v}} = \mathbf{b} - \mathbf{A}\hat{\mathbf{x}}$$

The 'least squares' solution can be found by varying the value of x until the following functional is minimized:

$$J(x) \equiv \sum_{i=1}^{m} v_i^2 = \mathbf{v}^T\mathbf{v} = (\mathbf{b} - \mathbf{A}\mathbf{x})^T (\mathbf{b} - \mathbf{A}\mathbf{x}).$$

That is, we are minimizing the sum of squares of the estimated residuals. If we vary $\mathbf{x}$ by a small amount, then $J(\mathbf{x})$ should also vary, except at the desired solution where it is stationary (since the slope of a function is zero at a minimum point). The following illustrates the application of this method to derive the least squares solution:

$$\delta J(\hat{\mathbf{x}}) = 0$$
$$\delta\{(\mathbf{b} - \mathbf{A}\hat{\mathbf{x}})^{\mathrm{T}}(\mathbf{b} - \mathbf{A}\hat{\mathbf{x}})\} = 0$$
$$\delta(\mathbf{b} - \mathbf{A}\hat{\mathbf{x}})^{\mathrm{T}}(\mathbf{b} - \mathbf{A}\hat{\mathbf{x}}) + (\mathbf{b} - \mathbf{A}\hat{\mathbf{x}})^{\mathrm{T}}\delta(\mathbf{b} - \mathbf{A}\hat{\mathbf{x}}) = 0$$
$$(-\mathbf{A}\delta\mathbf{x})^{\mathrm{T}}(\mathbf{b} - \mathbf{A}\hat{\mathbf{x}}) + (\mathbf{b} - \mathbf{A}\hat{\mathbf{x}})^{\mathrm{T}}(-\mathbf{A}\delta\hat{\mathbf{x}}) = 0$$
$$(-2\mathbf{A}\delta\mathbf{x})^{\mathrm{T}}(\mathbf{b} - \mathbf{A}\hat{\mathbf{x}}) = 0$$
$$(\delta\mathbf{x}^{\mathrm{T}}\mathbf{A}^{\mathrm{T}})(\mathbf{b} - \mathbf{A}\hat{\mathbf{x}}) = 0$$
$$\delta\mathbf{x}^{\mathrm{T}}(\mathbf{A}^{\mathrm{T}}\mathbf{b} - \mathbf{A}^{\mathrm{T}}\mathbf{A}\hat{\mathbf{x}}) = 0$$
$$\mathbf{A}^{\mathrm{T}}\mathbf{A}\hat{\mathbf{x}} = \mathbf{A}^{\mathrm{T}}\mathbf{b}$$

The last line is called the system of 'normal equations'. The solution to the normal equations is therefore:

$$\hat{\mathbf{x}} = (\mathbf{A}^{\mathrm{T}}\mathbf{A})^{-1}\mathbf{A}^{\mathrm{T}}\mathbf{b}$$

This assumes that the inverse to $\mathbf{A}^{\mathrm{T}}\mathbf{A}$ exists. For example, $m \geq 4$ is a necessary (but not sufficient) condition. Problems can exist if, for example, a pair of satellites lie in the same line of sight, or if the satellites are all in the same orbital plane. In almost all practical situations, $m \geq 5$ is sufficient. Alternatively, one parameter could be left unestimated (e.g., the height could be fixed to sea level for a boat).

3.2. Error Computation

The Covariance and Cofactor Matrices. If the observations $\mathbf{b}$ had no errors and if the model were perfect, then the estimates $\hat{\mathbf{x}}$ given by the above expression would be perfect. Any errors $\mathbf{v}$ in the original observations $\mathbf{b}$ will obviously map into errors $\mathbf{v}_x$ in the estimates $\hat{\mathbf{x}}$. It is also clear that this mapping will take exactly the same linear form as the above formula:

$$\mathbf{v}_x = (\mathbf{A}^{\mathrm{T}}\mathbf{A})^{-1}\mathbf{A}^{\mathrm{T}}\mathbf{v}$$

If we have (a priori) an expected value for the error in the data, σ, then we can compute the expected error in the parameters. We discuss the interpretation of the 'covariance matrix' later, but for now, we define it as the (square) matrix of expected values of one error multiplied by another error; that is, $C_{ij} \equiv \mathrm{E}(v_i v_j)$. A diagonal element C_{ii} is called a 'variance', and is often written as the square of the standard deviation, $C_{ii} = \mathrm{E}(v_i^2) = \sigma_i^2$. We can concisely define the covariance matrix by the following matrix equation:

$$\mathbf{C} \equiv \mathrm{E}(\mathbf{v}\mathbf{v}^{\mathrm{T}})$$

Let us for now assume we can characterize the error in the observations by one number, the variance $\sigma^2 = \mathrm{E}(v^2)$, which is assumed to apply to all m observations. Let us also assume that all observations are uncorrelated, $\mathrm{E}(v_i v_j) = 0$ (for $i \neq j$). We can therefore write the covariance matrix of observations as the diagonal matrix, $\mathbf{C}_\sigma = \sigma^2 \mathbf{I}$, where I is the $m \times m$ identity matrix:

$$C_\sigma = \begin{pmatrix} \sigma^2 & 0 & \cdots & 0 \\ 0 & \sigma^2 & & \vdots \\ \vdots & & \ddots & 0 \\ 0 & \cdots & 0 & \sigma^2 \end{pmatrix}_{m \times m}$$

Under these assumptions, the expected covariance in the parameters for the least squares solution takes a simple form:

$$C_x = E(v_x v_x^T)$$
$$= \sigma^2 (A^T A)^{-1}$$

Note that the 'cofactor matrix' $(A^T A)^{-1}$ also appears in the formula for the least squares estimate, $\hat{x}$. The 'cofactor matrix' is also sometimes called the 'covariance matrix', where it is implicitly understood that it should be scaled by the variance of the input observation errors. Since GPS observation errors are a strong function of the particular situation (e.g., due to environmental factors), it is common to focus on the cofactor matrix, which, like A, is purely a function of the satellite-receiver geometry at the times of the observations. The cofactor matrix can therefore be used to assess the relative strength of the observing geometry, and to quantify how the level of errors in the measurements can be related to the expected level of errors in the position estimates.

It should therefore be clear why A is called the 'design matrix'; we can in fact compute the cofactor matrix in advance of a surveying session if we know where the satellites will be (which we do, from the almanac in the Navigation Message). We can therefore 'design' our survey (in this specific case, select the time of day) to ensure that the position precision will not be limited by poor satellite geometry.

Interpreting the Covariance Matrix. The covariance matrix for the estimated parameters can be written in terms of its components:

$$C_x = \sigma^2 \left(A^T A \right)^{-1}$$

$$= \sigma^2 \begin{pmatrix} \sigma_x^2 & \sigma_{xy} & \sigma_{xz} & \sigma_{x\tau} \\ \sigma_{yx} & \sigma_y^2 & \sigma_{yz} & \sigma_{y\tau} \\ \sigma_{zx} & \sigma_{zy} & \sigma_z^2 & \sigma_{z\tau} \\ \sigma_{\tau x} & \sigma_{\tau y} & \sigma_{\tau z} & \sigma_\tau^2 \end{pmatrix}$$

As an example of how to interpret these components, if the observation errors were at the level of $\sigma = 1$ metre, the error in y coordinate would be at the level of σ_y metres; if the observation errors were $\sigma = 2$ metres, the error in y would be $2\sigma_y$ metres, and so on.

The off-diagonal elements indicate the degree of correlation between parameters. If σ_{yz} were negative, this means that a positive error in y will probably be accompanied by a negative error in z, and vice versa. A useful measure of correlation is the 'correlation coefficient', which is defined as

$$\rho_{ij} = \frac{\sigma_{ij}}{\sqrt{\sigma_i^2 \sigma_j^2}}$$

The correlation coefficient is only a function of the cofactor matrix, and is independent of the observation variance, σ^2. Its value can range between -1 to $+1$, where 0 indicates no correlation, and $+1$ indicates perfect correlation (i.e., the two parameters are effectively identical). Several textbooks show that the 'error ellipse' in the plane defined by the (z, y) coordinates (for example) can be computed using the elements σ_z^2, σ_y^2, and ρ_{zy}.

4.   The Carrier Phase Observable

4.1.   Concepts

We now introduce the carrier phase observable, which is used for high precision applications. We start with the basic concepts, starting with the meaning of 'phase', the principles of interferometry, and the Doppler effect. We then go on to describe the process of observing the carrier phase, and develop an observation model. Fortunately, most of the model can be reduced to what we have learned so far for the pseudorange. Unlike most textbooks, we take the approach of presenting the model in the 'range formulism', where the carrier phase is expressed in units of metres, rather than cycles. However, there are some fundamental differences between the carrier phase and the pseudorange observables, as we shall see when we discuss 'phase ambiguity' and the infamous problem of 'cycle slips'.

The Meaning of 'Phase', 'Frequency' and 'Clock Time'. 'Phase' is simply 'angle of rotation', which is conventionally in units of 'cycles' for GPS analysis. Consider a point moving anti-clockwise around the edge of a circle, and draw a line from the centre of the circle to the point. As illustrated in Figure 2.2, the 'phase' $\varphi(t)$ at any given time t can be defined as the angle through which this line has rotated.

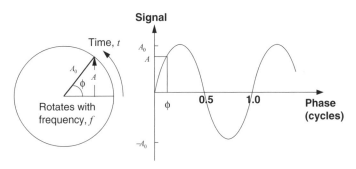

Figure 2.2   *The meaning of phase*

Phase is intimately connected with our concept of time, which is always based on some form of periodic motion, such as the rotation of the Earth, the orbit of the Earth around the Sun ('dynamic time'), or the oscillation of a quartz crystal in a wristwatch ('atomic time'). Even our reprentation of time is often based on rotation, such as the angle of the hands on the face of a clock. Angles of rotation give us our measure of 'time'. In this way, phase can be thought of as a measure of time (after conversion into appropriate units). We can write this formally as:

$$T(t) = k(\varphi(t) - \varphi_0)$$

where $T(t)$ is the time according to our clock at time t (whatever the clock may be), $\varphi_0 = \varphi(0)$ is so that the clock reads zero when $t = 0$, and k is a calibration constant, converting the units of cycles into units of seconds. Indeed, we can take the above equation as the *definition* of clock time. Whether or not this clock time is useful depends on the constancy of rate of change of phase. This brings us to the concept of frequency.

The 'frequency', expressed in units of 'cycles per second', is the number of times the line completes a full 360° rotation in one second (which of course, is generally a fractional number). This definition is somewhat lacking, since it seems to assume that the rotation is steady over the course of one second. One can better define frequency instantaneously as the first derivative of phase with respect to time, that is, the angular speed:

$$f \equiv \frac{d\varphi(t)}{dt}$$

We choose to treat phase as a fundamental quantity, and frequency as a derived quantity. For example, we can say that frequency is a constant, if we observe the phase as changing linearly in time. Constant frequency is the basis of an ideal clock. If the frequency can be written as a constant, f_0, then we can write the phase of an ideal clock as:

$$\varphi_{\text{ideal}} = f_0 t + \varphi_0$$

therefore:

$$T_{\text{ideal}} = k f_0 t$$

Since we want our clock second to equal a conventional second ($T_{\text{ideal}} = t$), we see that an appropriate choice for the calibration constant is $k = 1/f_0$, where f_0 is the nominal frequency of the oscillator. Going back to our original equation for clock time, we can now define clock time as:

$$T(t) = \frac{\varphi(t) - \varphi_0}{f_0}$$

How Phase Is Related to a Periodic Signal. At time t, the height of point $A(t)$ above the centre of the circle in Figure 2.2 is given by:

$$A(t) = A_0 \sin[2\pi\varphi(t)]$$

where A_0 is the radius of the circle. Since the concept of phase is often applied to periodic signals, we can call $A(t)$ the 'signal' and A_0 the 'amplitude of the signal'. For example, in the case of radio waves, $A(t)$ would be the strength of the electric field, which oscillates in time as the wave passes by. Inverting the above formula, we can therefore determine the phase $\varphi(t)$ if we measure the signal $A(t)$ (and, similarly, we could infer the clock time).

Note that, for an ideal clock, the signal would be a pure sinusoidal function of time:

$$
\begin{aligned}
A_{\text{ideal}} &= A_0 \sin 2\pi\varphi_{\text{ideal}} \\
&= A_0 \sin(2\pi f_0 t + 2\pi\varphi_0) \\
&= (A_0 \cos 2\pi\varphi_0)\sin 2\pi f_0 t + (A_0 \sin 2\pi\varphi_0)\cos 2\pi f_0 t \\
&= A_0^S \sin \omega_0 t + A_0^C \cos \omega_0 t
\end{aligned}
$$

where the 'angular frequency' $\omega_0 \equiv 2\pi f_0$ has units of radians per second. For a real clock, the signal would the same sinusoidal function of its own 'clock time', (but would generally be a complicated function of true time):

$$A(T) = A_0^S \sin \omega_0 T + A_0^C \cos \omega_0 T$$

We note that the nominal GPS signal takes on the above form, except that the signal is modulated by 'chips', formed by multiplying the amplitudes A_0^S (for C/A code) and A_0^C (for P code) by a pseudorandom sequence of +1 or −1. The underlying sinusoidal signal is called the 'carrier signal'. It is the phase of the carrier signal that gives us precise access to the satellite clock time; therefore we can use this phase for precise positioning.

Carrier Beat Signal. The GPS carrier signal $G(t)$ from the satellite is 'mixed' (multiplied) with the receiver's own replica carrier signal $R(t)$. The result of this mixing is shown in Figure 2.3.

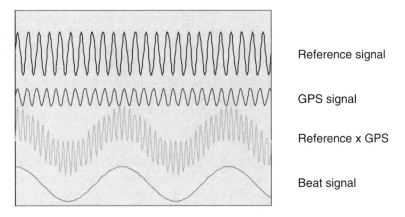

Reference signal

GPS signal

Reference x GPS

Beat signal

Figure 2.3 *Producing a beat signal by mixing the carrier and replica signals*

Mathematically, one can show that one would expect the result to be the difference between a low frequency signal and a high frequency signal:

$$R(t) \otimes G(t) = G_0 \sin 2\pi \varphi_G(t) \times R_0 \sin 2\pi \varphi_R(t)$$

$$= \frac{G_0 R_0}{2} [\cos 2\pi(\varphi_R(t) - \varphi_G(t)) - \cos 2\pi(\varphi_R(t) + \varphi_G(t))]$$

The high frequency component can easily be filtered out by the receiver electronics, leaving only the carrier beat signal:

$$B(t) = \text{Filter}\{R(t) \otimes G(t)\}$$

$$= \frac{G_0 R_0}{2} \cos 2\pi(\varphi_R(t) - \varphi_G(t))$$

$$\equiv B_0 \cos 2\pi(\varphi_B(t))$$

where we have introduced the carrier beat phase $\varphi_B(t)$, which is defined to be equal to the difference in phase between the replica signal and the GPS signal:

$$\varphi_B(t) \equiv \varphi_R(t) - \varphi_G(t)$$

By differentiating the above equation with respect to time, we find that the 'beat frequency' is equal to the difference in frequencies of the two input signals:

$$f_B \equiv \frac{d\varphi_B}{dt} = f_R - f_G$$

We note that the above formulas apply even when the carrier phase is modulated with codes, provided the replica signal is also modulated (because the values of −1 will cancel when multiplying the two signals). If the codes are not known, it is possible to square both the incoming signal and the replica signal prior to mixing. The problem with this is that squaring amplifies the noise, thus introducing larger random measurement errors.

Origin of the Phase Ambiguity. Our model of carrier beat phase is not a complete picture, since we can arbitrarily add an integer number of cycles to the carrier beat phase, and produce exactly the same observed beat signal. Suppose we only record the fractional phase of the first measurement. We would have no way of telling which integer N to add to this recorded phase so that it really did equal the difference in phase between the replica signal and the GPS signal. This is fundamentally because we have no direct measure of the total phase of the incoming GPS signal. We can express this as follows:

$$\Phi + N = \varphi_R - \varphi_G$$

where we use a capital Greek symbol Φ to emphasize that it represents the phase value actually recorded by the receiver. Provided the receiver does keep track of how many complete signal oscillations there have been since the first measurement, it can attach this number of cycles to the integer portion of the recorded beat phase. However, there will still be an overall ambiguity N that applies to all measurements. That is, we can model N as being the same (unknown) constant for all measurements. If the receiver loses count of the oscillations (e.g., because the signal is obstructed, or because of excessive noise), then a new integer parameter must be introduced to the model, starting at that time. This integer discontinuity in phase data is called a 'cycle slip'.

Interpretation of the Phase Ambiguity. The reader might also be wondering if there is some kind of geometrical interpretation for N. It turns out that there is, but it does require some oversimplified assumptions. By definition, the unknown value of N can be written as:

$$N = (\text{integer portion of } \varphi_R - \varphi_G) - (\text{integer portion of } \Phi)$$

The second term is completely arbitrary, and depends on the receiver firmware. For example, some receivers set this value to zero for the first measurement. Let us assume this is true, and drop this term. For the sake of interpretation, let us now assume that the receiver and satellite clocks keep perfect time. Under these circumstances, the first term would equal the integer portion of the number of signal oscillations that occur in the receiver from the time the signal was transmitted to the time the signal was received. We can therefore interpret N as equal to the number of carrier wavelengths between the receiver (at the time it makes the first observation), and the satellite (at the time it transmitted the signal). Of course, we made assumptions about perfect clocks and the particular nature of the firmware; so we must beware not to take this interpretation too literally.

Intuitive Model: the Doppler Effect. How can phase be used to measure distance? One way hinted at above is that the phase essentially tells you the clock time. As we shall see in the next section, we can develop phase in almost the same way as the pseudorange model. Another intuitive way of looking at it is to consider the Doppler Effect. We are all familiar with the acoustic version of the Doppler Effect, as we hear a vehicle's engine at a higher pitch when it is approaching, and a lower pitch when receding. Can we use the Doppler Effect to design a distance-measuring device?

Imagine two perfect clocks; one is at a fixed point, the other is approaching in a vehicle. Let both clocks be generating a sinusoidal signal. The frequency difference between the reference signal, and the approaching signal, increases with the vehicle's speed of approach. Let us build a receiver to mix the two signals and measure the beat signal. The beat frequency would be a measure of the speed.

Let us count the cycles of the beat signal; or better yet, let us measure the phase (cycles plus fractional cycles) of the beat signal. Clearly, the beat phase would measure the change in distance to vehicle. We can therefore (after appropriate unit conversion) write the intuitive equation:

$$\text{Beat phase} = \text{distance to vehicle} + \text{constant}$$

This demonstrates that, although the beat phase can be used to precisely measure change in distance from one time to another, there is an unknown constant which prevents us from knowing the full distance. This can be seen by considering moving the reference observer 10 metres away from the original position, and then repeating the experiment. The Doppler effect is clearly exactly the same, and the number of cycles passing by would not change. The very first value of the measured beat phase will indeed be different, but this single measurement cannot be used to infer distance. For example, we have already stated that we don't know what integer number of cycles to attribute to the first beat phase measurement.

4.2.   Carrier Phase Observation Model

Carrier Beat Phase Model. We now move towards a more rigorous treatment of the carrier beat phase observable, building on our concepts of phase and signal mixing. Our notation will change slightly in preparation for further development. To summarize what we know already, the satellite carrier signal (from antenna) is mixed with a reference signal generated by receiver's clock. The result, after high pass filtering, is a 'beating' signal. The phase of this beating signal equals the reference phase minus the incoming GPS carrier phase from a satellite; however, it is ambiguous by an integer number of cycles. From this point on, 'carrier beat phase' will be simply called 'carrier phase' (but it should not be confused with the phase of the incoming signal!).

Observation of satellite S produces the carrier phase observable Φ^S:

$$\Phi^S(T) = \varphi(T) - \varphi^S(T) - N^S$$

where φ is the replica phase generated by the receiver clock, and φ^S is the incoming signal phase received from GPS satellite S. The measurement is made when the receiver clock time is T. Now take the point of view that the phase of the incoming signal received at receiver clock time T is identical to the phase that was transmitted from the satellite at satellite clock time T^S:

$$\varphi^S(x, y, z, T) = \varphi^S_{transmit}(x^S, y^S, z^S, T^S_{transmit})$$

Of course, if we adopt this point of view, then we shall eventually have to consider the model of how long it takes a wavefront of constant phase to propagate from the satellite to the receiver, so that we may model the appropriate satellite clock time at the time of signal transmission, T^S. We shall return to that later.

As discussed previously, we can write clock time as a function of phase and nominal frequency:

$$T(t) = \frac{\varphi(t) - \varphi_0}{f_0}$$

We can therefore substitute all the phase terms with clock times:

$$\varphi(T) = f_0 T + \varphi_0$$
$$\varphi^S_{transmit}(T^S) = f_0 T^S_{transmit} + \varphi^S_0$$

Therefore, the carrier phase observable becomes:

$$\Phi^S(T) = f_0 T + \varphi_0 - f_0 T^S - \varphi_0^S - N^S$$
$$= f_0(T - T^S) + \varphi_0 - \varphi_0^S - N^S$$

where we implicitly understand that the clock times refer to different events (reception and transmission, respectively).

We note that terms containing the superscript S are different for each satellite, but all other terms are identical. Receivers are designed and calibrated so that the phase constant φ_0 is identical for all satellites; that is, there should be no inter-channel biases. Receivers should also sample the carrier phase measurements from all satellites at exactly the same time. (If the receivers have multiplexing electronics to save on cost, then the output should have been interpolated to the same epoch for all satellites.) The time T^S will vary slightly from satellite to satellite, since the satellite transmission time must have been different for all signals to arrive at the same time. We also note that the last three terms are constant, and cannot be separated from each other. We can collectively call these terms the 'carrier phase bias', which is clearly not an integer.

In preparation for multi-receiver and multi-satellite analysis, we now introduce the subscripts A, B, C, etc. to indicate quantities specific to receivers, and we introduce superscripts j, k, l, etc. to identify satellite-specific quantities. We write the carrier phase observed by receiver A from satellite j:

$$\Phi_A^j(T_A) = f_0(T_A - T^j) + \varphi_{0A} - \varphi_0^j - N_A^j$$

Note that data should be sampled at exactly the same values of clock time (called 'epochs') for all receivers, so all values of T_A are identical at a given epoch. However, receivers, clocks do not all run at exactly the same rate, therefore the true time of measurement will differ slightly from receiver to receiver. Also, note that each receiver–satellite pair has a different carrier phase ambiguity.

Range Formulation. It is convenient to convert the carrier phase model into units of range. This simplifies concepts, models, and software. In the range formulation, we multiply the carrier phase equation by the nominal wavelength.

$$L_A^j(T_A) \equiv \lambda_0 \Phi_A^j(T_A)$$
$$= \lambda_0 f_0(T_A - T^j) + \lambda_0(\varphi_{0A} - \varphi_0^j - N_A^j)$$
$$= c(T_A - T^j) + \lambda_0(\varphi_{0A} - \varphi_0^j - N_A^j)$$
$$\equiv c(T_A - T^j) + B_A^j$$

where we still retain the name 'carrier phase' for $L_A^j(T_A)$, which is in units of metres. We see immediately that this equation is identical to that for the pseudo-range, with the exception of the 'carrier phase bias', B_A^j which can be written (in units of metres):

$$B_A^j \equiv \lambda_0(\varphi_{0A} - \varphi_0^j - N_A^j)$$

Note that the carrier phase bias for (undifferenced) data is not an integer number of wavelengths, but also includes unknown instrumental phase offsets in the satellite and receiver.

We have not yet mentioned any differences between carrier phase on the L1 and L2 channel. Although they have different frequencies, in units of range the above equations take the same form. Actually, the clock bias parameters would be identical for both L1 and L2 phases, but the carrier phase bias would be different. The main difference comes when we develop the model in terms of the propagation delay, which is a function of frequency in the Earth's ionosphere.

Observation Model. We note that the first term in the carrier phase model is simply the pseudorange, and the second term is a constant. We have already developed a simplified model for pseudorange, so we can therefore write a model for carrier phase as follows:

$$
\begin{aligned}
L_A^j(T_A) &= c(T_A - T^j) + B_A^j \\
&= \rho_A^j(t_A, t^j) + c\tau_A - c\tau^j + Z_A^j - I_A^j + B_A^j
\end{aligned}
$$

In the above expression, we have explicitly included the delay on the signal due to the troposphere Z_A^j and the ionosphere $-I_A^j$ (the minus sign indicating that the phase velocity actually increases). Models for the atmospheric delay terms are beyond the scope of this text.

The model for pseudorange can be similarly improved, with the small difference that the ionospheric delay has a positive sign:

$$
\begin{aligned}
P_A^j(T_A) &= c(T_A - T^j) \\
&= \rho_A^j(t_A, t^j) + c\tau_A - c\tau^j + Z_A^j + I_A^j
\end{aligned}
$$

This is because, from physics theory, any information, such as the +1 and −1 'chips' which are modulated onto the carrier wave, must travel with the 'group velocity' rather than 'phase velocity'. According to the theory of relativity, information can not be transmitted faster than c. From the physics of wave propagation in the ionosphere, it can be shown that the group delay is (to a very good first-order approximation) precisely the same magnitude, but opposite sign of the phase delay (which is really a phase 'advance').

Accounting for Time-Tag Bias. Before proceeding, we return to the problem posed in our discussion of the pseudorange model; that is, we typically do not know the true time of signal reception t_A which we need to calculate the satellite-receiver range term $\rho_A^j(t_A, t^j)$ precisely. From section 3.1, the true time of reception can be written:

$$
t_A = T_A - \tau_A
$$

where the epoch T_A is known exactly, as it is the receiver clock time written into the data file with the observation (and hence called the 'time-tag'). However, the receiver clock bias τ_A is not known initially, but could be as large as milliseconds. The problem is that, due to satellite motion and Earth rotation, the range will change by several metres over the period of a few milliseconds, so we must be careful to account for this for precision work (especially when using the carrier phase observable). For precision work (1 mm), we should use a value τ_A that is accurate to $1\,\mu s$.

There are various approaches to dealing with this in GPS geodetic software, which typically use some combination of the following methods:

1 Use values of the receiver clock bias computed in a first step using a pseudo-range point position solution at each epoch.
2 Iterate the least squares procedure, processing both carrier phase and pseudo-range data simultaneously, and using estimates of the clock bias to compute the true receive time, and therefore the new range model.
3 Use an estimate $\hat{t}^j$ of the true transmit time t^j to compute the satellite position:

$$\hat{t}^j = \hat{T}^j - \tau^j$$
$$= (T_A - P_A^j/c) + \tau^j$$

where the satellite clock bias τ^j is obtained from the Navigation Message. One can then directly compute the range term and true receive time with sufficient precision, provided the approximate station coordinates are known to within 300 m (corresponding to the 1 μs timing requirement). Interestingly, this is the basis for 'time transfer', since it allows one to compute the receiver clock bias using pseudorange data from only one GPS satellite. As a method for computing range for precise positioning, this is not often used, perhaps for the reason that it is not a pure model, as it depends on pseudorange data and approximate positions.

4 One can take a modelling 'short cut' to avoid iteration by expanding the range model as a first-order Taylor series. Since this method often appears in the development of the observation equation in textbooks, we discuss it in more detail here.

A Note on the Range-Rate Term. The observation equation can be approximated as follows:

$$L_A^j(T_A) = \rho_A^j(t_A, t^j) + c\tau_A - c\tau^j + Z_A^j - I_A^j + B_A^j$$
$$= \rho_A^j(T_A - \tau_A, t^j) + c\tau_A - c\tau^j + Z_A^j - I_A^j + B_A^j$$
$$\approx \rho_A^j(T_A, t^{j\prime}) - \dot{\rho}_A^j\tau_A + c\tau_A - c\tau^j + Z_A^j - I_A^j + B_A^j$$
$$= \rho_A^j(T_A, t^{j\prime}) + (c - \dot{\rho}_A^j)\tau_A - c\tau^j + Z_A^j - I_A^j + B_A^j$$

where we see that the effect can be accounted for by introducing the modelled range rate (i.e., the relative speed of the satellite in the direction of view). The 'prime' for the satellite transmit time $t^{j\prime}$ (which is used to compute the satellite coordinates) is to indicate that it is not the true transmit time, but the time computed using the nominal receive time T_A. A first-order Taylor expansion has been used. The higher order terms will only become significant error sources if the receiver clock bias is greater than about 10 ms, which does not usually happen with modern receivers. In any case, clock biases greater than this amount would result in a worse error in relative position due to the effect of S/A.

Textbooks sometimes include a 'range rate' term in the development of the phase observation model, even though, strictly speaking, it is unnecessary. After all, the first line of the above equation is correct, and the lack of a priori knowledge of the receiver clock bias can easily be dealt with by least squares iteration, or prior point positioning using the pseudorange. On the other hand, it is nevertheless instructional to show the above set of equations, since it does illustrate that it is more correct to use $(c - \dot{\rho}_A^j)$ as the partial derivatives with respect to the receiver clock in the design matrix, rather than simply using c. This is crucial if one is not initializing clocks using point position solutions or iteration (as is typical, for example, with the GIPSY OASIS II software). It is not important if initialization of τ_A is achieved with $1\,\mu s$ accuracy.

In the expressions to follow, we shall not explicitly include the range rate term on the assumption that time-tag bias has been handled one way or another.

4.3.   Differencing Techniques

Single Differencing. The purpose of 'single differencing' is to eliminate satellite clock bias. Consider the observation equations for two receivers, A and B observing the same satellite, j:

$$L_A^j = \rho_A^j + c\tau_A - c\tau^j + Z_A^j - I_A^j + B_A^j$$
$$L_B^j = \rho_B^j + c\tau_B - c\tau^j + Z_B^j - I_B^j + B_B^j$$

The single difference phase is defined as the difference between these two:

$$\Delta L_{AB}^j \equiv L_A^j - L_B^j$$
$$= (\rho_A^j + c\tau_A - c\tau^j + Z_A^j - I_A^j + B_A^j) - (\rho_B^j + c\tau_B - c\tau^j + Z_B^j - I_B^j + B_B^j)$$
$$= (\rho_A^j - \rho_B^j) + (c\tau_A - c\tau_B) - (c\tau^j - c\tau^j) + (Z_A^j - Z_B^j) - (I_A^j - I_B^j) - (B_A^j - B_B^j)$$
$$= \Delta\rho_{AB}^j + c\Delta\tau_{AB} + \Delta Z_{AB}^j - \Delta I_{AB}^j + \Delta B_{AB}^j$$

where we use the double subscript to denote quantities identified with two receivers, and the triangular symbol as a mnemonic device, to emphasize that the difference is made between two points on the ground. The geometry of single differencing is illustrated in Figure 2.4.

An assumption has been made that the satellite clock bias τ^j is effectively identical at the slightly different times that the signal was transmitted to A and to B.

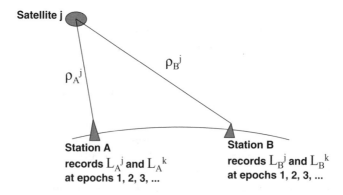

Satellite j

ρ_A^j ρ_B^j

Station A
records L_A^j and L_A^k
at epochs 1, 2, 3, ...

Station B
records L_B^j and L_B^k
at epochs 1, 2, 3, ...

Figure 2.4   *Single differencing geometry*

The difference in transmission time could be as much as a few milliseconds, either because the imperfect receiver clocks have drifted away from GPS time by that amount, or because the stations might be separated by 1000 km or more. Since selective availability is typically at the level of 10^{-9} (variation in frequency divided by nominal frequency), over a millisecond (10^{-3} s) the satellite clock error will differ by 10^{-12} s. This translates into a distance error of $10^{-12}c$, or 0.3 mm, a negligible amount under typical S/A conditions (however, it may not be negligible if the level of S/A were increased; but this effect could in principle be corrected if we used reference receivers to monitor S/A). Another point worth mentioning is that the coordinates of the satellite at transmission time can easily be significantly different for receivers A and B, and this should be remembered when computing the term $\Delta\rho_{AB}^{j}$.

The atmospheric delay terms are now considerably reduced, and vanish in the limit that the receivers are standing side by side. The differential troposphere can usually be ignored for horizontal separations less than approximately 30 km, however, differences in height should be modelled. The differential ionosphere can usually be ignored for separations of 1 to 30 km, depending on ionospheric conditions. Due to ionospheric uncertainty, it is wise to calibrate for the ionosphere using dual-frequency receivers for distances greater than a few kilometres.

Although the single difference has the advantage that many error sources are eliminated or reduced, the disadvantage is that only relative position can be estimated (unless the network is global-scale). Moreover, the receiver clock bias is still unknown, and very unpredictable. This takes us to 'double differencing'.

Double Differencing. The purpose of 'double differencing' is to eliminate receiver clock bias. Consider the single differenced observation equations for two receivers A and B observing satellites j and k:

$$\Delta L_{AB}^{j} = \Delta\rho_{AB}^{j} + c\Delta\tau_{AB} + \Delta Z_{AB}^{j} - \Delta I_{AB}^{j} + \Delta B_{AB}^{j}$$
$$\Delta L_{AB}^{k} = \Delta\rho_{AB}^{k} + c\Delta\tau_{AB} + \Delta Z_{AB}^{k} - \Delta I_{AB}^{k} + \Delta B_{AB}^{k}$$

The double difference phase is defined as the difference between these two:

$$\begin{aligned}
\nabla\Delta L_{AB}^{jk} &\equiv \Delta L_{AB}^{j} - \Delta L_{AB}^{k} \\
&= (\Delta\rho_{AB}^{j} + c\Delta\tau_{AB} + \Delta Z_{AB}^{j} - \Delta I_{AB}^{j} + \Delta B_{AB}^{j}) \\
&\quad - (\Delta\rho_{AB}^{k} + c\Delta\tau_{AB} + \Delta Z_{AB}^{k} - \Delta I_{AB}^{k} + \Delta B_{AB}^{k}) \\
&= (\Delta\rho_{AB}^{j} - \Delta\rho_{AB}^{k}) + (c\Delta\tau_{AB} - c\Delta\tau_{AB}) + (\Delta Z_{AB}^{j} - \Delta Z_{AB}^{k}) - (\Delta I_{AB}^{j} - \Delta I_{AB}^{k}) \\
&\quad - (\Delta B_{AB}^{j} - \Delta B_{AB}^{k}) \\
&= \nabla\Delta\rho_{AB}^{jk} + \nabla\Delta Z_{AB}^{jk} - \nabla\Delta I_{AB}^{jk} + \nabla\Delta B_{AB}^{jk}
\end{aligned}$$

where we use the double-superscript to denote quantities identified with two satellites, and the upside-down triangular symbol as a mnemonic device, to emphasize that the difference is made between two points in the sky. Figure 2.5 illustrates the geometry of double differencing.

One point worth mentioning is that although the receiver clock error has been eliminated to first order, the residual effect due 'time tag bias' on the computa-

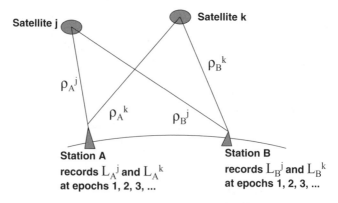

Figure 2.5   *Double differencing geometry*

tion of the range term does not completely disappear, and still needs to be dealt with if the receiver separation is large.

Any systematic effects due to unmodelled atmospheric errors are generally increased slightly by approximately 40% by double differencing as compared to single differencing. Similarly, random errors due to measurement noise and multipath are increased. Overall, random errors are effectively doubled as compared with the undifferenced observation equation. On the other hand, the motivation for double differencing is to remove clock bias, which would create much larger errors.

One could process undifferenced or single differenced data, and estimate clock biases. In the limit that clock biases are estimated at every epoch (the 'white noise clock model') these methods become almost identical, provided a proper treatment is made of the data covariance (described later). It is almost, but not quite identical, because differencing schemes almost always involve preselection of baselines in a network to form single differences, and data can be lost by lack of complete overlap of the observations to each satellite. (This problem can be minimized by selecting the shortest baselines in the network to process, and by assuring that no more than one baseline be drawn to a receiver with a significant loss of data.)

Double Differenced Ambiguity.   The double difference combination has an additional advantage, in that the ambiguity is an integer:

$$\begin{aligned}
\nabla\Delta B_{AB}^{jk} &= \Delta B_{AB}^{j} - \Delta B_{AB}^{k}\\
&= (B_A^{j} - B_B^{j}) - (B_A^{k} - B_B^{k})\\
&= \lambda_0\,(\varphi_{0A} - \varphi_0^{j} - N_A^{j}) - \lambda_0\,(\varphi_{0B} - \varphi_0^{j} - N_B^{j}) - \lambda_0\,(\varphi_{0A} - \varphi_0^{k} - N_A^{k})\\
&\quad + \lambda_0\,(\varphi_{0B} - \varphi_0^{k} - N_B^{k})\\
&= -\lambda_0\,(N_A^{j} - N_B^{j} - N_A^{k} + N_B^{k})\\
&= -\lambda_0\nabla\Delta N_{AB}^{jk}
\end{aligned}$$

Hence we can write the double differenced phase observation equation:

$$\nabla\Delta L_{AB}^{jk} = \nabla\Delta\rho_{AB}^{jk} + \nabla\Delta Z_{AB}^{jk} - \nabla\Delta I_{AB}^{jk} - \lambda_0\nabla\Delta N_{AB}^{jk}$$

From the point of view of estimation, it makes no difference whether we use a minus or plus sign for N, so long as the partial derivative has a consistent sign (which, for the above equation, would be $-\lambda_0$).

Triple Differencing. The purpose of 'triple differencing' is to eliminate the integer ambiguity. Consider two successive epochs $(i, i + 1)$ of double differenced data from receivers A and B observing satellites j and k:

$$\nabla\Delta L_{AB}^{jk}(i) = \nabla\Delta\rho_{AB}^{jk}(i) + \nabla\Delta Z_{AB}^{jk}(i) - \nabla\Delta I_{AB}^{jk}(i) - \lambda_0\nabla\Delta N_{AB}^{jk}$$
$$\nabla\Delta L_{AB}^{jk}(i + 1) = \nabla\Delta\rho_{AB}^{jk}(i + 1) + \nabla\Delta Z_{AB}^{jk}(i + 1) - \nabla\Delta I_{AB}^{jk}(i + 1) - \lambda_0\nabla\Delta N_{AB}^{jk}$$

The triple difference phase is defined as the difference between these two:

$$\delta(i, i + 1)\nabla\Delta L_{AB}^{jk} \equiv \nabla\Delta L_{AB}^{jk}(i + 1) - \nabla\Delta L_{AB}^{jk}(i)$$
$$= \delta(i, i + 1)\nabla\Delta\rho_{AB}^{jk}(i) + \delta(i, i + 1)\nabla\Delta Z_{AB}^{jk}(i) - \delta(i, i + 1)\nabla\Delta I_{AB}^{jk}(i)$$

where we use the delta symbol to indicate the operator that differences data between epochs. Figure 2.6 illustrates triple differencing geometry.

The triple difference only removes the ambiguity if it has not changed during the time interval between epochs. Any cycle slips will appear as outliers, and can easily be removed by conventional techniques. This is unlike the situation with double differencing, where cycle slips appear as step functions in the time series of data.

The disadvantage of the triple difference is that it introduces correlations between observations in time. Generally, increasing correlations in data has the property of decreasing the data weights. With triple differencing, the degradation in precision is substantial, so triple differenced data are inappropriate for precise surveys. On the other hand, it is a very useful method for obtaining better nominal parameters for double differencing (to ensure linearity), and it is a

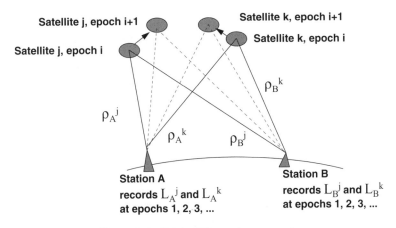

Figure 2.6 Triple differencing geometry

robust method, due to the ease with which cycle slips can be identified and removed.

It can be shown that triple difference solution is identical to the double differenced solution, provided just one epoch double differenced equation is included for the first point in a data arc, along with the triple differences, and provided the full data covariance matrix is used to compute the weight matrix. This special approach can provide tremendous savings in computation time over straightforward double differencing, while retaining robustness.

5.   Relative Positioning Using Carrier Phase

5.1.   Selection of Observations

Linear Dependence of Observations. We can usually form many more possible combinations of double differenced observations than there are original data. This poses a paradox, since we cannot create more information than we started with. The paradox is resolved if we realize that some double differences can be formed by differencing pairs of other double differences. It then becomes obvious that we should not process such observations, otherwise we would be processing the same data more than once. This would clearly be incorrect.

Figure 2.7 illustrates the simplest example of the problem. In this example, we have three satellites j, k and l, observed by two receivers A and B. If we ignore trivial examples (e.g., $L_{AB}^{jk} = -L_{AB}^{kj}$), there are three possible double differences that can be formed:

$$L_{AB}^{jk} = (L_A^j - L_B^j) - (L_A^k - L_B^k)$$
$$L_{AB}^{jl} = (L_A^j - L_B^j) - (L_A^l - L_B^l)$$
$$L_{AB}^{lk} = (L_A^l - L_B^l) - (L_A^k - L_B^k)$$

Note that we can form any one of these observations as a linear combination of the others:

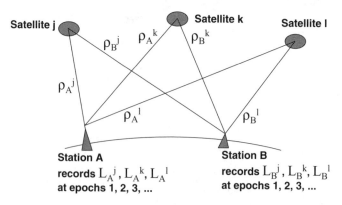

Figure 2.7   *Double difference geometry with three satellites*

$$L_{AB}^{jk} = L_{AB}^{jl} + L_{AB}^{lk}$$
$$L_{AB}^{jl} = L_{AB}^{jk} - L_{AB}^{lk}$$
$$L_{AB}^{lk} = L_{AB}^{jk} - L_{AB}^{jl}$$

The data set $\{L_{AB}^{jk}, L_{AB}^{jl}, L_{AB}^{lk}\}$ is therefore said to be linearly dependent. A linearly independent set must be used for least squares. Examples of appropriate linearly independent sets in this example are:

$$\{L_{AB}^{jk}, L_{AB}^{jl}\} = \Lambda^j \equiv \{L_{AB}^{ab}|a = j; b \neq j\}$$
$$\{L_{AB}^{kj}, L_{AB}^{kl}\} = \Lambda^k \equiv \{L_{AB}^{ab}|a = k; b \neq k\}$$
$$\{L_{AB}^{lj}, L_{AB}^{lk}\} = \Lambda^l \equiv \{L_{AB}^{ab}|a = l; b \neq l\}$$

The Reference Satellite Concept. The 'reference satellite concept' involves using either set Λ^j, Λ^k or Λ^l throughout the data set. For example, double differences in set Λ^l all involve the satellite l. All sets are equally valid, and will produce identical solutions provided the data covariance is properly constructed (see the next section). Obviously, the reference satellite itself has to have data at every epoch, otherwise data will be lost. This can cause problems for less sophisticated software. Typically, a reference satellite should be picked which has the longest period in view. A better algorithm is to select a reference satellite epoch by epoch.

Our simple example can be easily extended to more than three satellites. For example, consider satellites 1, 2, 3, 4 and 5 in view. We can pick satellite 4 as the reference satellite, therefore, our linearly independent set is:

$$\Lambda^4 \equiv \{L_{AB}^{ab}|a = 4; b \neq 4\}$$
$$= \{L_{AB}^{41}, L_{AB}^{42}, L_{AB}^{43}, L_{AB}^{45}\}$$

Note that for a single baseline (i.e. two receivers), the number of linearly independent double differenced observations is $s - 1$, where s is the number of satellites being tracked.

The Reference Station Concept. However, if we have a network of more than two receivers, we must account for the fact that double differenced data from the set of all baselines is linearly dependent. We therefore introduce the 'reference station' concept, where our double differences all include a common reference station. This guarantees linear independence. For example, consider satellites 1, 2, 3 and 4 being tracked by stations A, B, and C. If we pick our reference satellite to be 3, and reference station to be B, then our chosen set is:

$$\Lambda_B^3 \equiv \{L_{cd}^{ab}|a = 3; b \neq 3; c = B, d \neq B\}$$
$$= \{L_{BA}^{31}, L_{BA}^{32}, L_{BA}^{34}, L_{BC}^{31}, L_{BC}^{32}, L_{BC}^{34}\}$$

Note that the number of linearly independent double differenced observations is $(s - 1)(r - 1)$, where s is the number of satellites being tracked, and r is the number of receivers. So, in our previous example, three receivers and four satellites gives six observations. This assumes that s satellites are observed by all stations. This may not be the case, either due to obstructions, receiver problems, or because the receivers are separated by such a large distance that the satellite is not above the horizon for some receivers.

If using the reference station concept, it is therefore best to choose a receiver close to the middle of a large network, with few obstructions, and no hardware problems, otherwise the set of double differences may not be as complete as it could be. The reference station concept is obviously not optimal, and is seriously problematic for large networks. A better strategy for large networks is to select short baselines that connect together throughout the entire network, being careful not to introduce linear dependent observations, by not including any closed polygons (such as triangles) in the network. In principle, there must be only one possible path between pairs of stations. An even better strategy would be to optimize this choice for every epoch.

Solution Uniqueness. It should be stressed that, if all stations were tracking the same set of satellites at all epochs, then the selection of reference station and reference satellite would not matter, since an identical solution would be produced, whatever the selection. This assumes that the data weight matrix is properly constructed (as described below) and that no data outliers are removed.

The problem of linear dependence usually introduces a level of arbitrariness into the solutions due to violation of the above assumptions. This problem is also true even if the previously suggested improvements are made to the reference station concept, since the user typically has to make decisions on which baselines to process (even for more sophisticated software). This is somewhat unsatisfactory, since there is generally no unique solution, However, experience shows that any reasonable selection will only produce small differences in the final solutions.

There is a way to produce a unique solution, and that is to process undifferenced observations, estimating clock parameters at each epoch. As stated previously, this will produce a solution identical to double differencing under ideal conditions. This class of software is not typically available commercially; however, it should be stressed that double differencing software does not produce significantly inferior results for most situations. What is far more important is the quality of the observable models, the range of appropriate estimation options, and the ability to detect and deal with cycle slips and outliers.

5.2.   Baseline Solution Using Double Differences

Simplified Observation Equations. We now show how relative coordinates can be estimated between two receivers using the double differenced carrier phase data. We start by simplifying the observation equation, assuming that the relative atmospheric delay is negligible for short distances between receivers. We also drop the symbols '$\nabla\Delta$' of the previous section to simplify the notation. We shall therefore use the following simplified observation equation:

$$L_{AB}^{jk} = \rho_{AB}^{jk} - \lambda_0 N_{AB}^{jk}$$

General Procedure. Processing double differenced data from two receivers results in a 'baseline solution'. The estimated parameters include the vector between the two receivers, in Cartesian coordinates $(\Delta x, \Delta y, \Delta z)$ and may include parameters to model the tropospheric delay. In addition, the ambiguity parameters N_{AB}^{jk} for each set of double differences to specific satellite pairs (j, k) must be estimated.

The observation equations therefore require linearization in terms of all these parameters (according to the process explained in Section 4.1). Typically, one station is held fixed at good nominal coordinates, which quite often come from an initial pseudorange point position solution. We should mention, however, that due to S/A, point position solutions can have substantial errors (100 m) which may create significant errors in the double differenced observation model, and in the design matrix.

If we call the fixed station A, then estimating the baseline vector is equivalent to estimating the coordinates of station B. It is convenient to formulate the problem to estimate parameters (x_B, y_B, z_B). For example, consider a GPS survey between stations A and B, which observe satellites 1, 2, 3 and 4 for every epoch in the session, where we arbitrarily pick satellite 2 as the reference satellite. For every epoch i, we have the following linearly independent set of three double differenced observations:

$$\Lambda^2(i) \equiv \{L_{AB}^{ab}(i) | a = 2; b \neq 2\}$$
$$= \{L_{AB}^{21}(i), L_{AB}^{23}(i), L_{AB}^{24}(i)\}$$

We therefore have the parameter set $(x_B, y_B, z_B, N_{AB}^{21}, N_{AB}^{23}, N_{AB}^{24})$. If any cycle slips had occurred and could not be corrected, then additional ambiguity parameters must be added to the list.

As for point positioning, the linearized observation equations can be expressed in the form:

$$\mathbf{b} = \mathbf{Ax} + \mathbf{v}$$

where the residual observations are listed in the b matrix, which has dimensions $d \times 1$, where d is the number of linearly independent double differenced data. The design matrix A has dimensions $d \times p$ where p is the number of parameters, and the parameter corrections are contained in the x matrix, which has dimensions $p \times 1$. The observation errors are represented by the v matrix, which has the same dimensionality as $\mathbf{b}$. We shall discuss the design matrix later on.

It is important to use a 'weighted least squares' approach, because of correlations in the double differenced data. We shall not derive the weighted least squares estimator here, but for completeness, the solution is given here:

$$\hat{\mathbf{x}} = (\mathbf{A}^T \mathbf{WA})^{-1} \mathbf{A}^T \mathbf{Wb}$$

where $\mathbf{W}$ is the data weight matrix, to be derived later on, and $\mathbf{b}$ is a vector containing the double-differenced residual observations. See Cross (1994) for a full derivation of the weighted least squares estimator.

The covariance matrix for the estimated parameters is given by:

$$\mathbf{C_x} = (\mathbf{A^T W A})^{-1}$$

The covariance matrix can be used to judge whether the theoretically expected precision from the observation scenario is sufficient to allow ambiguities to be fixed to integer values. If ambiguity parameters can be fixed in the model, a theoretically more precise solution can be generated from the same data, but without estimating the ambiguities. This process will necessarily reduce the covariance matrix, lowering the expected errors in the station coordinates. This does not necessarily mean that the solution is better, but that it statistically ought to be better, assuming the integers were correctly fixed. The assessment of solution accuracy goes beyond the scope of this discussion, but basically one can compare results with previous results (using GPS, or even some other technique). In addition, how well the data is fit by the model is reflected in the standard deviation of the post-fit residuals.

The Design Matrix. The coefficients of the design matrix can be illustrated by looking at a single row, for example, corresponding to observation $L_{AB}^{24}(i)$:

$$A_{AB}^{24}(i) = \left(\frac{\partial L_{AB}^{24}(i)}{\partial x_B} \quad \frac{\partial L_{AB}^{24}(i)}{\partial y_B} \quad \frac{\partial L_{AB}^{24}(i)}{\partial z_B} \quad \frac{\partial L_{AB}^{24}(i)}{\partial N_{AB}^{21}} \quad \frac{\partial L_{AB}^{24}(i)}{\partial N_{AB}^{23}} \quad \frac{\partial L_{AB}^{24}(i)}{\partial N_{AB}^{24}} \right)$$

$$= \left(\frac{\partial \rho_{AB}^{24}(i)}{\partial x_B} \quad \frac{\partial \rho_{AB}^{24}(i)}{\partial y_B} \quad \frac{\partial \rho_{AB}^{24}(i)}{\partial z_B} \quad 0 \quad 0 \quad -\lambda_0 \right)$$

As an example of one of the partial derivatives for one of the coordinates:

$$\frac{\partial \rho_{AB}^{24}(i)}{\partial x_B} = \frac{\partial}{\partial x_B}\left(\rho_A^2(i) - \rho_B^2(i) - \rho_A^4(i) + \rho_B^4(i) \right)$$

$$= \frac{\partial \rho_A^2(i)}{\partial x_B} - \frac{\partial \rho_B^2(i)}{\partial x_B} - \frac{\partial \rho_A^4(i)}{\partial x_B} + \frac{\partial \rho_B^4(i)}{\partial x_B}$$

$$= \frac{\partial \rho_B^4(i)}{\partial x_B} - \frac{\partial \rho_B^2(i)}{\partial x_B}$$

$$= \frac{x_{B0} - x^4(i)}{\rho_B^4(i)} - \frac{x_{B0} - x^2(i)}{\rho_B^2(i)}$$

Minimum Data Requirements for Least Squares. For a least squares solution, a necessary condition is that the number of data must exceed the number of estimated parameters:

$$d \geq p$$

where we allow for the 'perfect fit solution' $(d = p)$. Under the assumption that all receivers track the same satellites for every epoch, the number of linearly independent double differences is

$$d = q(r - 1)(s - 1)$$

where q is the number of epochs, r the number of receivers, and s is the number of satellites. Assuming no cycle-slip parameters:

$$p = 3 + (r - 1)(s - 1)$$

where there are $(r - 1)(s - 1)$ ambiguity parameters. Therefore,

$$q(r - 1)(s - 1) \geq 3 + (r - 1)(s - 1)$$
$$(q - 1)(r - 1)(s - 1) \geq 3$$

Now, we know that $s \geq 2$ and $r \geq 2$ for us form double differences. Therefore, we can deduce that $q \geq 4$ if we have the minimal geometry of two receivers and two satellites (only one double difference per epoch!). Obviously, this minimal configuration is very poor geometrically, and would not be recommended as a method of precise positioning.

Note that no matter how many receivers or satellites we have, q is an integer, and therefore under any circumstance, we must have at least $q \geq 2$. That is, we cannot do single epoch relative positioning, if we are estimating integer ambiguities. If we can find out the ambiguities by some other means, then single epoch relative positioning is possible. Otherwise, we have to wait for the satellite geometry to change sufficiently in order to produce a precise solution.

For a single baseline $r = 2$ with 2 epochs of data $q = 2$ (which we should assume are significantly separated in time), the minimum number of satellites to produce a solution gives the condition $s \geq 4$. Interestingly, this corresponds to the minimum number of satellites for point positioning. If a tropospheric parameter were also being estimated, the condition would be $s \geq 5$. Of course, these conditions can be relaxed if we have more than 2 epochs, however, it is the end-points of a data arc which are most significant, since they usually represent the maximum geometrical change which we require for a good solution. In summary, one can achieve very good results over short distances with only four satellites, but over longer distances where the troposphere must be estimated, a minimum of five satellites is recommended.

5.3. Stochastic Model

Statistical Dependence of Double Differences. We have seen how double differences can be linearly dependent. The problem we now address is that double differenced observations that involve a common receiver and common satellite are statistically dependent. For example, at a given epoch, double differences L_{AB}^{21}, L_{AB}^{23} and L_{AB}^{24} are correlated due to the single differenced data in common, L_{AB}^{2}. Any measurement error in this single difference will contribute exactly the same error to each of the double differences. Therefore, a positive error in L_{AB}^{21} is statistically more likely to be accompanied by a positive error in L_{AB}^{23}. As another example, if we are processing a network using a reference satellite j and reference receiver A, all double differences in the linearly independent set will be statistically dependent because of the data in common, L_{A}^{j}.

Data Weight Matrix for Double Differences. In a situation where data is correlated, weighted least squares is appropriate. To complete our description of how to compute a relative position estimate, we therefore need to explain how to compute the appropriate data weight matrix, **W**. The construction of **W** can be generally called the 'stochastic model', which describes the statistical nature of our data (as opposed to the 'functional model' described so far, from which the observables can be computed deterministically).

(As an aside for more advanced readers, some software process undifferenced observations, estimating clock biases as 'stochastic parameters' at every epoch. It should be emphasized that there is an equivalence between explicit estimation of 'stochastic parameters', and the use of an appropriate 'stochastic model' which, in effect, accounts for the missing parameters through the introduction of correlations in the data. In principle, any parameter can either be estimated as part of the functional model, or equivalently removed using an appropriate stochastic model. To go more into this is beyond the scope of this text.)

The weight matrix is the inverse of the covariance matrix for the double differenced data:

$$\mathbf{W} = \mathbf{C}_{\nabla\Delta}^{-1}$$

which has dimensions $q(r-1)(s-1) \times q(r-1)(s-1)$.

We start by assuming a covariance matrix for undifferenced data (i.e., the actually recorded data), which has dimensions $qrs \times qrs$. Typically, this is assumed to be diagonal, since the receiver independently measures the signals from each satellite separately. We shall, however, keep the form general. So the problem is, given a covariance matrix for undifferenced data, how do we compute the covariance matrix for double differenced data? This is achieved using the rule of propagation of errors, which we will see in Chapter 3, where geocentric coordinates are mapped into topocentric coordinates using an affine transformation. By analogy, we can deduce that the covariance of double-differenced data can be written:

$$\mathbf{C}_{\nabla\Delta} = \mathbf{D}\mathbf{C}\mathbf{D}^{\mathbf{T}}$$

where **D** is the matrix which transforms a column vector of the recorded data into a column vector of double differenced data:

$$\nabla\Delta\mathbf{L} = \mathbf{D}\mathbf{L}$$

Clearly, **D** is a rectangular matrix with the number of rows equal to the number of linearly independent double differenced data, and the number of columns equal to the number of recorded data. Using our previous assumptions, **D** has dimensions $q(r-1)(s-1) \times qrs$. The components of **D** must have values of +1, −1, or 0, arranged so that we produce a linearly independent set of double differences. To complete this discussion, the double differenced data weight matrix can be written:

$$\mathbf{W} = (\mathbf{D}\mathbf{C}\mathbf{D}^{\mathbf{T}})^{-1}$$

Covariance Matrix for Estimated Parameters. As we have already seen, for weighted least squares we can write the computed covariance matrix for estimated parameters as:

$$\mathbf{C_x} = (\mathbf{A^T W A})^{-1}$$

We can now write down the full expression for the computed covariance matrix, by substituting for the double differenced data weight matrix $\mathbf{W}$:

$$\mathbf{C_x} = (\mathbf{A^T (D C D^T)^{-1} A})^{-1}$$

As mentioned above, for the (undifferenced) data covariance $\mathbf{C}$, we often use a diagonal matrix, assuming a value for the standard deviation of an observation. Typical realistic values for this are several mm. Although the receiver can usually measure the phase with better precision than a mm, the post-fit residuals typically show several mm standard deviations, due to unmodelled errors such as multipath.

Even using such an inflated value for measurement precision might not produce a realistic covariance matrix for station coordinates. This is partly due to two reasons: (1) unmodelled errors can be correlated with the parameters being estimated (an 'aliasing effect'); and (2) post-fit almost always show some degree of time-correlation (e.g., due to multipath). A simple and often surprisingly effective way to deal with this problem is to multiply the final coordinate covariance matrix by an empirical scaling factor, inferred 'by experience', according to the brand of software being used, the observation scenario, and the estimation strategy used.

6. Introducing High Precision GPS Geodesy

6.1. High Precision Software

The observable model discussed so far has been very basic, as it glosses over advanced features that are important for high precision software. Several software packages have been developed since the 1980s that are capable of delivering high precision geodetic estimates over long baselines. This software is a result of intensive geodetic research, mainly by universities and government research laboratories.

Typical features of such software include:

- orbit integration with appropriate force models;
- accurate observation model (Earth model, media delay . . .) with rigorous treatment of celestial and terrestrial reference systems;
- reliable data editing (cycle-slips, outliers);
- estimation of all coordinates, orbits, tropospheric bias, receiver clock bias, polar motion, and Earth spin rate;
- ambiguity resolution algorithms applicable to long baselines;

- estimation of reference frame transformation parameters and kinematic modelling of station positions to account for plate tectonics and co-seismic displacements.

We can summarize the typical quality of geodetic results from 24 hours of data:

- relative positioning at the level of few parts per billion of baseline length;
- absolute (global) positioning at the level of 1 cm in the IERS Terrestrial Reference Frame (ITRF);
- tropospheric delay estimated to a few mm;
- GPS orbits determined to 10 cm;
- Earth pole position determined to 1 cm;
- clock synchronization (relative bias estimation) to 100 ps.

Two features of commercial software are sometimes conspicuously absent from more advanced packages: (1) sometimes double differencing is not implemented, but instead, undifferenced data is processed, and clock biases are estimated; and (2) network adjustment using baseline solutions is unnecessary, since advanced packages do a rigorous, one-step, simultaneous adjustment of station coordinates directly from all available GPS observations.

Some precise software packages incorporate a Kalman filter (or an equivalent formulism). This allows certain selected parameters to vary in time, according to a statistical ('stochastic') model. Typically this is used for the tropospheric bias, which can vary as a random walk in time. A filter can also be used to estimate clock biases, where 'white noise' estimation of clock bias approaches the theoretical equivalent of double differencing.

Although many more packages have been developed, there are three ultra high-precision software packages which are widely used around the world by researchers and are commonly referenced in the scientific literature:

1 BERNESE software, developed by the Astronomical Institute, University of Bern, Switzerland.
2 GAMIT software, developed by the Massachusetts Institute of Technology, USA.
3 GIPSY software, developed by the Jet Propulsion Laboratory, California Institute of Technology, USA.

There are several other packages, but they tend to be limited to the institutions that wrote them. It should be noted that, unlike commercial software packages, use of the above software can require a considerable investment in time to understand the software and how best to use it under various circumstances. Expert training is often recommended by the distributors.

6.2. Sources of Data and Information

For high precision work, it is important to abide by international reference system standards and use the best available sources of data and ancillary information. We therefore summarize two especially important international sources

of data information for the convenience of the interested reader:
Features of the IERS (International Earth Rotation and Reference Frame Service) are:

- Central Bureau;
- documented IERS Conventions for observation models and reference systems;
- IERS Annual Reports;
- IERS Terrestrial Reference Frame for reference station coordinates;
- routine publication of Earth rotation parameters.

Features of the IGS (International GNSS Service) are:

- Central Bureau located at the Jet Propulsion Laboratory, USA;
- documented IGS Standards for permanent GPS stations;
- oversees operation of global GPS network (~300 stations);
- distributes tracking data and precise ephemerides;
- maintains on-line database with Internet access.

7. Conclusion

Having read and understood this chapter, you should now understand the basics of GPS positioning observation models and parameter estimation. You should also have an appreciation of the difference between basic positioning, and the more advanced positioning using high precision software packages. You should now have good background knowledge and an appropriate context to prepare you for more advanced material!

3

Datum Transformations and Projections

This chapter describes the nature of the Earth in relation to mass and gravity, the ellipsoid, the Earth's surface, the geoid and mean sea level (MSL). Also described is the relationship between orthometric heights and GPS heights with the requirement, and related procedures, to transform GPS height to mean sea level height, such as Ordnance Datum Newlyn height. Projections are briefly discussed, which are required to transfer measurements and positions from the ellipsoid onto a flat surface suitable for making into a map.

1. Integration Requirements

Integration of GPS and GIS requires a common coordinate system to describe the positions of objects. The problem is that GPS produces coordinates in a global Cartesian coordinate system, whereas for many GIS applications (such as road navigation) it is more appropriate that the coordinates relate to standard maps. Quite often the user interface to a GPS-GIS integrated system will involve a map on a flat surface. For many applications it is often simpler to implement algorithms in the GIS that directly use regional map coordinates that relate to a local flat projection. The specific projection used in regional maps will depend on the location of the user. Standard map projections often vary between countries and states. The thing that ties all these various local systems together is the WGS-84 global reference system used by GPS. National surveys and mapping agencies provide information that allow users (typically through software) to transform coordinates from WGS-84 to maps, and vice versa. In an integrated GPS-GIS system this transformation is typically transparent to the user. Quite

Intelligent Positioning: GIS-GPS Unification G. Taylor and G. Blewitt
© 2006 John Wiley & Sons, Ltd

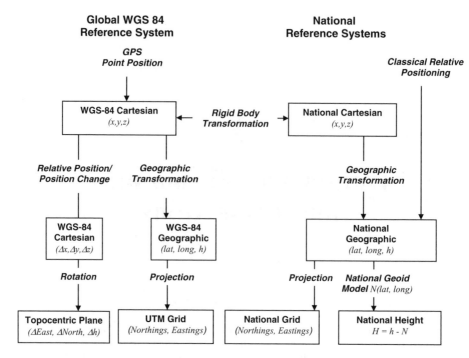

Figure 3.1   *Transforming GPS coordinates into useful map coordinates*

often the user only wishes to know his or her position in relation to other objects, such as city streets.

In Chapter 2 we explained how GPS is used to position a receiver in a global Cartesian reference system. In this chapter we explain more about this global reference system, and how it relates to national or state map projections. First, we explain how coordinates generally relate to the physical shape and gravity field of the Earth. Then we show the fundamental principles involved in converting coordinates from one system into another (Figure 3.1). In the following sections each of the components of Figure 3.1 will be defined in more detail, and the relationship between each system and the physical Earth will be explained.

2.   Global Reference Systems

2.1.   WGS-84 Cartesian Coordinates

GPS positions a user in the same reference frame as the GPS satellites, 'World Geodetic System' WGS-84. The satellite orbits are in turn estimated using data from the US Department of Defense's GPS tracking stations around the world. It is the defined coordinates of these tracking stations that 'realize' (make real) the WGS-84 reference frame. In effect, the GPS satellites act as the mediator between the user and these fundamental stations, thus giving the user access to the WGS-84 frame anywhere in the world.

The tracking station positions are specified by three Cartesian coordinates in units of the S.I. metre. It is believed that the internal accuracy of these coordinates is now approximately 1 cm. Here 'internal accuracy' refers to the level of self-consistency of these coordinates with respect to the true physical distances and internal angles of the station network geometry.

WGS-84 is an example of an Earth-fixed, geocentric terrestrial reference system. Ideally the frame origin $(0, 0, 0)$ lies at the centre of the Earth, with the z-axis of unit vector $(0, 0, 1)$ pointing towards the North Pole, the x-axis $(1, 0, 0)$ towards the equator on the Prime Meridian (historically, the 'Greenwich Meridian'), and the y-axis defined so as to complete the right-handed triplet (on the equator pointing at $90°$ longitude). The axes rotate in inertial space along with the solid Earth, so the axes appear to be 'fixed' in the solid Earth. The 'external accuracy' of WGS-84 refers to its alignment with the ideal WGS-84. It is believed that external accuracy is also now at the 1 cm level.

2.2.   International Terrestrial Reference System (ITRS)

WGS-84 was initially established when the prototype GPS system was being tested in the 1980s. At that time the internal accuracy of WGS-84 was believed to be at the level of decimetres, with external accuracy at the one metre level. Since the 1990s the U.S. Department of Defense has gradually improved the accuracy of WGS-84 by bringing it into agreement with the International Terrestrial Reference Frame (ITRF).

The International Earth Rotation Service (IERS) was established in 1988 by the international community and was charged with producing a terrestrial reference system to support the stringent requirements of space geodesy, the science of the Earth's changing shape and gravity field, and its application to geodynamics, tectonics, and other areas of Earth sciences. The origin of the International Terrestrial Reference System (ITRS) is defined as located at the centre of mass of the entire Earth system, including the oceans and atmosphere. The z-axis is defined to align with the North Pole on a specific reference date, since the pole position is known to move about from day to day, a phenomenon known as 'polar motion'. The x-axis points to the intersection of the equator and the Prime Meridian, however, note that the physical marker of zero longitude at Greenwich no longer defines the Prime Meridian. In fact, it lies a few hundred metres away. Besides the arbitrariness in the definition of Prime Meridian, one problem is that the Earth's crust actually moves everywhere due to plate tectonics. The Prime Meridian is now defined so that it does not rotate on average with respect to the Earth's surface, which means that it is continually drifting with respect to any specific mark on the Earth's surface, including the historical marker at Greenwich. The ITRS takes into account many effects such as polar motion, tectonics and tidal deformation, which is a necessary requirement to assign meaningful coordinates with few millimetres accuracy.

The specific reference frame ITRF2000 lists the coordinates of hundreds of stations around the globe from a mix of space geodetic techniques, including GPS, very long baseline interferometry (VLBI), satellite laser ranging (SLR), and the French Doppler ranging satellite system DORIS. Since the 1990s, the interoperability between WGS-84 and ITRF has meant that, effectively, the resolution of WGS-84 has been extended to the few-millimetre level. This level of accuracy is achievable by users with the right hardware, the right software, and the right input data. Access to ITRF by GPS users is possible through the International GNSS Service (IGS), which has been officially operational since 1994. IGS operates a similar but far more extensive global tracking network than the U.S. Department of Defense. Funded internationally by various space agencies, national geodetic agencies, research institutions, and universities, the IGS currently operates a growing network of about 300 stations distributed around the globe. These stations have precise ITRF coordinates, and are used to determine the GPS satellite positions with much higher accuracy than is available from the Broadcast Message. Distributed over the Internet, these orbit positions and satellite clock parameters are available in varying degrees of accuracy depending on their latency. Real-time products are available to position users to within 10 cm. Non-real-time users can position themselves to within a few millimetres, using the type of precise positioning software described in Chapter 2.

2.3.   WGS-84 Ellipsoidal Coordinates

So far we have emphasized the fundamental nature of Cartesian coordinates used in the computation of a user's position using GPS, and in the definition of global terrestrial reference frames. It is very important to realize that when using GPS, other systems are entirely secondary to the Cartesian system. That is, a station's latitude and longitude derive from its Cartesian coordinates, and as such, latitude and longitude are secondary concepts. However, it is intuitively more useful for many applications to use a latitude/longitude coordinate system.

WGS-84 defines an ellipsoid that is designed to closely fit the actual shape of the Earth. Of course, the Earth is not perfectly ellipsoidal, however, the ellipsoid allows for a relatively simple and well-specified procedure to define the longitude, latitude, and height of any point on or near the Earth. Note that the Earth's gravity field only approximates an ellipsoid, so the meaning of 'height' will be slightly different from its traditional meaning, since the ellipsoid is not a truly 'level' surface. For this reason, height defined in this manner is called 'ellipsoidal height' to differentiate it from 'orthometric height', which can be thought of as 'height above sea level'.

The centre of the ellipsoid is at the Cartesian origin. In accordance with the definition of ITRS, this is at the centre of mass of the entire Earth system, and so has physical significance. The shape of an ellipsoid is defined through three semi-major axes. For all ellipsoids used for mapping, there are only two semi-major axes, because the equatorial radius is defined to be constant. The polar radius is smaller than the equatorial radius, hence the ellipsoid is said to be

Table 3.1   *WGS-84 ellipsoidal parameters*

Parameter	Value
Equatorial radius, a	6378.135 m
Polar radius, $b = a(1 - 1/f)$	6355.135 m
Flattening factor, $f = a/(a - b)$	298.12345

oblate. Table 3.1. shows the parameters associated with WGS-84 ellipsoid. Note that the polar radius is actually derived from the defined value of the 'flattening factor'.

WGS-84 reflects the physical shape of the Earth in that the polar radius is approximately 23 km shorter than the equatorial radius. Although not part of the formal definition of WGS-84, it can be shown that the mean radius of the Earth is approximately 6371 km. To an extremely good approximation, the surface area and volume of the ellipsoid can be computed using the usual formulas for a perfect sphere of this mean radius.

If we were to define 'the geoid' as the level surface (surface of constant gravitational potential) that best fits the sea surface, then WGS-84 approximates the geoid, but differs in some regions by as much as approximately 100 metres. Clearly, therefore, 'ellipsoidal height' in the WGS-84 system may be quite different from height above sea level. It is quite understandable that countries have historically defined their own ellipsoidal systems so that the ellipsoidal surface closely fits sea level in that country (an approximation that would generally break down quickly with distance from that country). Note that continental topography has a range of approximately 10 km, which is 100 times larger than the sea level discrepancy. So from this point of view, the WGS-84 ellipsoid does remarkably well as an appropriate reference surface, considering that it is a global compromise.

2.4.   Cartesian to Ellipsoidal Transformation

Cartesian coordinates (x, y, z) can, to an approximation better than 0.1 mm, be converted into ellipsoidal coordinates of latitude, longitude, and height (ϕ, λ, h) using the following formulas:

$$\phi = \arctan\left[\frac{z}{\left(x^2 + y^2\right)^{1/2}}\right]$$

$$\lambda = \arctan\left[\frac{y}{x}\right]$$

$$h = \left(x^2 + y^2 + z^2\right)^{1/2} - a$$

Note that care must taken when using the arctan function to apply appropriate logic to the signs of y and x so that the answer for longitude lies in the correct quadrant. In spreadsheet software this logic is often conveniently taken care of by the function 'arctan2(x,y)'. This is most important for longitude. For latitude, the result must of course lie between $-90°$ to $+90°$, and so the quadrant is always determined appropriately by the sign of z.

2.5.   Ellipsoidal to Cartesian Transformation

If ellipsoidal coordinates are given in some database from another source, it is straightforward to convert these to Cartesian coordinates provided the correct ellipsoidal parameters (a, f) are understood, or are provided in the metadata:

$$x = (a+h)\cos\phi\cos\lambda$$
$$y = (a+h)\cos\phi\sin\lambda$$
$$z = (a+h)\sin\phi$$

2.6.   Relative Coordinates: Cartesian to Topocentric

A change in position from time A to time B, or the position of object B relative to object A, can be described by a vector of WGS-84 'geocentric' coordinate displacements:

$$\Delta x = x_B - x_A$$
$$\Delta y = y_B - y_A$$
$$\Delta z = z_B - z_A$$

This vector can also be expressed by 'topocentric coordinates' $(\Delta e, \Delta n, \Delta h)$, which relate to displacements in the local East, North, and ellipsoidal height directions. This is achieved by a rotation $\mathbf{G}$ defined by the following matrix equation:

$$\begin{pmatrix} \Delta e \\ \Delta n \\ \Delta h \end{pmatrix} = \mathbf{G} \begin{pmatrix} \Delta x \\ \Delta y \\ \Delta z \end{pmatrix}$$

$$\mathbf{G} = \begin{pmatrix} -\sin\lambda & \cos\lambda & 0 \\ -\sin\phi\cos\lambda & -\sin\phi\sin\lambda & \cos\phi \\ \cos\phi\cos\lambda & \cos\phi\sin\lambda & \sin\phi \end{pmatrix}$$

Displacements or relative positions expressed in topocentric coordinates can be converted into geocentric coordinates, using the orthogonal property of rotations:

$$\begin{pmatrix} \Delta x \\ \Delta y \\ \Delta z \end{pmatrix} = \mathbf{G}^{-1} \begin{pmatrix} \Delta e \\ \Delta n \\ \Delta h \end{pmatrix}$$

$$\mathbf{G}^{-1} = \mathbf{G}^T = \begin{pmatrix} -\sin\lambda & -\sin\phi\cos\lambda & \cos\phi\cos\lambda \\ \cos\lambda & -\sin\phi\sin\lambda & \cos\phi\sin\lambda \\ 0 & \cos\phi & \sin\phi \end{pmatrix}$$

Note that topocentric coordinates so defined still relate to the WGS-84 system, and so the directions of East, North and height will generally not be aligned with a specific national system. However, the alignment will typically be to within 10^{-5} radians, so within a local area of 1 km, relative topocentric coordinates will agree with the national system to within a few centimetres. For the change in the position of a car from one second to the next, relative topocentric coordinates will agree with the national system to better than one millimetre, and so vehicle velocity can always be adequately expressed in the WGS-84 system no matter which national system is used. For many applications within a local context, and for many navigational algorithms, it is therefore appropriate to use WGS-84 topocentric coordinates to describe relative positions, change in position, velocities, object orientation, differences in height, etc., because WGS-84 is compatible with all national systems at such a local scale.

2.7.   GPS Estimated Errors: Cartesian to Topocentric

In Section 3.2 of Chapter 2 we discussed computation of the covariance matrix, which describes the statistically expected level of errors in the estimated Cartesian coordinates. Here we show how to transform the error estimates into topocentric coordinates. This turns out to be crucial later in the book for road navigation algorithms to rule out highly improbable solutions. One cannot assume that errors in Cartesian geocentric coordinates are a reasonable measure of the error in topocentric coordinates, because quite frequently the error in height is far worse than in the East and North directions. Fortunately, the computation is quite straightforward, as it follows from the law of propagation of errors.

First of all, note that a position error can be represented as a small displacement, and therefore it transforms according to:

$$\mathbf{v}_{env} = \mathbf{G}\mathbf{v}_{xyz}$$

where the geocentric coordinate matrix $\mathbf{v}_{xyz} = (\Delta x, \Delta y, \Delta z)^T$ and the topocentric coordinate matrix $\mathbf{v}_{env} = (\Delta e, \Delta n, \Delta h)^T$.

We now derive how to transform the covariance matrix of coordinates from the geocentric system to the topocentric system:

$$
\begin{aligned}
\mathbf{C}_{env} &= E\left(\mathbf{v}_{env}\mathbf{v}_{env}^T\right) \\
&= E\left((\mathbf{G}\mathbf{v}_{xyz})(\mathbf{G}\mathbf{v}_{xyz})^T\right) \\
&= E\left(\mathbf{G}\mathbf{v}_{xyz}\mathbf{v}_{xyz}^T\mathbf{G}^T\right) \\
&= \mathbf{G}E\left(\mathbf{v}_{xyz}\mathbf{v}_{xyz}^T\right)\mathbf{G}^T \\
&= \mathbf{G}\mathbf{C}_{xyz}\mathbf{G}^T
\end{aligned}
$$

where $\mathbf{C}_{xyz}$ is really the 3×3 submatrix taken from the original 4×4 matrix (which also included coefficients for the clock parameter τ). The covariance matrix in the topocentric system $\mathbf{C}_{env}$ can be written in terms of its components:

$$\mathbf{C}_{env} = \sigma^2 \begin{pmatrix} \sigma_e^2 & \sigma_{en} & \sigma_{eh} \\ \sigma_{en} & \sigma_n^2 & \sigma_{nh} \\ \sigma_{eh} & \sigma_{nh} & \sigma_h^2 \end{pmatrix}$$

We could then use this covariance, for example, to plot error ellipses in the horizontal plane.

2.8.   Dilution of Precision

We can now define the various types of 'dilution of precision' (DOP) as a function of diagonal elements of the covariance matrix in the topocentric system:

$$VDOP \equiv \sigma_h$$
$$HDOP \equiv \sqrt{\sigma_e^2 + \sigma_n^2}$$
$$PDOP \equiv \sqrt{\sigma_e^2 + \sigma_n^2 + \sigma_h^2}$$

where, for example, $VDOP$ stands for 'vertical dilution of precision', H stands for horizontal, P for position. As an example of how to interpret DOP, a standard deviation of 1 metre in observations would give a standard deviation in horizontal position of $HDOP$ metres. If $VDOP$ had a value of 5, we could expect pseudorange errors of 1 metre to map into vertical position errors of 5 metres, and so on.

According to the equations, the cofactor matrix and therefore the DOP values are purely a function of satellite geometry as observed by the receiver. A 'good geometry' therefore gives low DOP values. A 'bad geometry' can give very high DOP values. As a general rule, $PDOP$ values larger than 5 are considered poor. If there are fewer than a sufficient number of satellites to produce a solution, or if 2 out of 4 satellites lie in approximately the same direction in the sky, then the cofactor matrix becomes singular, and the DOP values go to infinity. The above formulas assume that all four parameters (x, y, z, τ) are being estimated. If fewer parameters are estimated, for example, if height is not estimated because we are in a boat at sea level, or if we have a digital topographic map, then the modified DOP values would get smaller, and they would no longer generally be infinity for only three satellites in view.

For the same reasons as given in Section 2.6, the estimated errors in the local topographic system and the associated DOP values are valid not only in WGS-84, but also for regional reference systems. This is because errors represent such a short displacement, that any slight misalignment of the two reference systems is of insignificant consequence.

3.   Regional Reference Systems

3.1.   Regional Ellipsoidal Cordinates

By regional reference systems, we include national systems, state systems, and even international regional systems. Examples include the North American

Datum NAD-83 in the United States and Canada, and the Ordnance Survey's OSBG-36 in Great Britain. Each system is defined in terms of a regional ellipsoid that is typically selected to approximate sea level. Longitude and latitude are defined with some level of arbitrariness by assigning coordinates to at least one 'fundamental station', which said to fix the 'datum' so that longitude and latitude are unambiguously defined.

There is no reason to expect that regional latitude and longitude will be even close to WGS-84 coordinates or between different regional systems. In the United Kingdom, for example, the difference between OSGB-36 and WGS-84 amounts to several hundred metres. In North America, NAD-83 is an exception, as it does align at the metre scale with WGS-84 because it was designed that way. NAD-83 is adopted by the US National Geodetic Survey (NGS) and Natural Resources Canada (NRCan). Another difference is the treatment of plate tectonics. In ITRF (equivalent to WGS-84) station coordinates are specified at a reference date together with a velocity to account for plate tectonics and other types of crustal motion. In many regional systems, station coordinates are assumed to be stationary, so that station velocity does not form part of the reference frame. This implies that coordinates in a regional frame generally change in time with respect to WGS-84 coordinates, as that region is transported over the globe by plate tectonics.

Crustal deformation presents a similar problem. In Europe, most of Europe behaves as a stable tectonic block, and so this portion of Europe is used to define a stable frame. Since all of California is deforming, the State of California together with NGS have defined a smooth reference station velocity field as part of the California reference system. It should be emphasized here that the real need for regional frames is to do with legal issues and infrastructure needs. Geophysicists who are actually studying deformation of the Earth through plate tectonics and other processes do not use such legally defined regional systems, which are in any case typically insufficient to meet the more stringent requirements of science. The methods used by geophysicists are not discussed here, other than to note that their work forms the foundation for the modern realization of WGS-84 through the ITRF, thus making possible millimetre-level coordinate grid resolution.

Table 3.2 (Dutch 2004) lists various regional systems, their ellipsoidal parameters, and method of projection which will be explained later. How do we transform ellipsoidal coordinates from one reference system into another? The answer is to first transform ellipsoidal coordinate of the first system into Cartesian coordinates (Section 2). Second, the Cartesian coordinates undergo a 'similarity' or 'rigid-body' transformation, on the assumption that each system is physically internally accurate, and that differences between two systems are due to orientation, definition of the origin (datum), and perhaps the scale. The third step is to convert the transformed Cartesian coordinates into ellipsoidal coordinates using the appropriate ellipsoidal parameters of the second system (Section 1).

The rigid-body transformation is defined as follows, in going from coordinates in system A to coordinates in system B:

Table 3.2   *Ellipsoidal to ellipsoidal coordinates*

Datum	Equatorial Radius, metres (a)	Polar Radius, metres (b)	Flattening (a − b)/a	Use
WGS-84/ NAD-83	6,378,137	6,356,752.3142	1/297.257223563	Global/North America
GRS 80	6,378,137	6,356,752.3141	1/297.257222101	Global (scientific)
WGS72	6,378,135	6,356,750.5	1/298.26	DOD (obsolete)
Australian 1965	6,378,160	6,356,774.7	1/298.25	Australia
Krasovsky 1940	6,378,245	6,356,863.0	1/298.3	Soviet Union
International (1924) – Hayford (1909)	6,378,388	6,356,911.9	1/297	Global (obsolete)
Clake 1880	6,378,249.1	6,356,514.9	1/293.46	France, Africa
Clarke 1866	6,378,206.4	6,356,583.8	1/294.98	North America
Airy 1830	6,377,563.4	6,356,256.9	1/299.32	Great Britain
Bessel 1841	6,377,397.2	6,356,079.0	1/299.15	Central Europe, Chile, Indonesia
Everest 1830	6,377,276.3	6,356,075.4	1/300.80	South Asia

$$\begin{pmatrix} x \\ y \\ z \end{pmatrix}_B = \begin{pmatrix} 1+s & \theta_z & -\theta_y \\ -\theta_z & 1+s & \theta_x \\ \theta_y & -\theta_x & 1+s \end{pmatrix} \begin{pmatrix} x \\ y \\ z \end{pmatrix}_A + \begin{pmatrix} T_x \\ T_y \\ T_z \end{pmatrix}$$

where the seven parameters are: s is scale increment; θ_x, θ_y, and θ_z are small angles of rotation about the x, y, and z axes; and T_x, T_y, and T_z account for a translation between the origins of the two systems. Typical values for scale and angles (in radians) are 10^{-5} or smaller. Typical values for the translation are <1 km.

Since WGS-84 is nominally now attached to ITRF, it is quite generally the case that the small angles of rotation between WGS-84 and regional systems change from year to year as plate tectonics gradually rotates the underlying plate to which the regional reference system is attached. This change is very small for many applications, typically 10^{-8} per year, which translates to decimetres over a decade.

In practice, the seven parameters from WGS-84 to national systems have been estimated by governmental agencies such as the NGS (US) and the Ordnance Survey (UK), and such values are usually made available to the general public. Indeed, GPS firmware or processing software often store this information so that results can be converted into a regional system of the user's choice.

The above transformation equation only strictly applies if each of the two reference frames are accurate. National systems are quite often historically based, and for legal reasons can retain significant physical errors for sake of continuity.

This is the case in the UK, for example, where the physical distance between old triangulation pillars at opposite ends of the country can be in error at the metre level. This is caused by regional distortions in the network. To deal with this problem, the UK publishes transformation parameters for different regions of its network. A recommended procedure ensures that coordinates do not jump discontinuously as the user moves from one region to another.

Each country has had to deal with historical problems one way or the other, and thus it is impossible to specify here a general method that is applicable everywhere. One approach that we predict will eventually happen in all countries is a national frame will be adopted that is fundamentally defined in terms of ITRF, but adjusted only to mitigate the average effect of plate tectonic rotation. This is the approach that has been taken in Europe. The main obstacles to this are no longer technical, but are of a legal nature. As national surveys increasingly use GPS and future GNSS systems such as Galileo, the use of ITRF (= WGS-84) as the fundamental reference frame will gain gradual acceptance to the point that governments will likely begin to accept this new, efficient, and accurate way of spatially referencing its infrastructure. Then it will only be a matter of time before the old systems disappear.

3.2.   Plane Coordinates

Each country typically specifies a different method for projecting longitude and latitude onto a flat map. In large countries the method of projection for regional maps, such as state maps, generally depend on the region's location. The most common method for projection is the Universal Transverse Mercator projection (UTM), or something similar.

One thing to keep in mind is that all projections necessarily distort the true three-dimensional surface of the ellipsoid when mapping onto a flat plane. Thus, the actual map that is rendered to the user is merely a convenient interface, and it is in no way fundamental to the way GPS-GIS integrated systems work or 'think'. On the contrary, GPS-GIS integrated systems can use the tools described above to represent the spatial relationship between objects and the user in a truly three-dimensional way.

3.3.   Converting Latitude and Longitude to UTM

These formulas are taken from Dutch (2004) which are slightly modified from Army (1973). They are accurate to within less than a metre within a given grid zone.

Symbols

lat   = latitude of point
long   = longitude of point
$long_0$ = central meridian of zone
k_0 = scale along $long_0$ = 0.9996

e $= \text{SQRT}(1 - b^2/a^2) = 0.08$ approximately. This is the eccentricity of the Earth's elliptical cross-section.

e'^2 $= (ea/b)^2 = e^2/(1 - e^2) = 0.007$ approximately. The quantity e' only occurs in even powers so it need only be calculated as e'^2.

n $= (a - b)/(a + b)$

rho $= a(1 - e^2)/(1 - e^2\sin^2(\text{lat}))^{3/2}$. This is the radius of curvature of the Earth in the meridian plane.

nu $= a/(1 - e^2\sin^2(\text{lat}))^{1/2}$. This is the radius of curvature of the earth perpendicular to the meridian plane. It is also the distance from the point in question to the polar axis, measured perpendicular to the earth's surface.

p $= (\text{long} - \text{long}_0)$

$\sin 1''$ = sine of one second of arc $= \text{pi}/(180*60*60) = 4.8481368 \times 10^{-6}$.

Calculate the Meridional Arc. S is the meridional arc through the point in question (the distance along the earth's surface from the equator). All angles are in radians:

S $= A'\text{lat} - B'\sin(2\text{lat}) + C'\sin(4\text{lat}) - D'\sin(6\text{lat}) + E'\sin(8\text{lat})$, where lat is in radians

$A' = a[1 - n + (5/4)(n^2 - n^3) + (81/64)(n^4 - n^5) \ldots]$
$B' = (3an/2)[1 - n + (7/8)(n^2 - n^3) + (55/64)(n^4 - n^5) \ldots]$
$C' = (15an^2/16)[1 - n + (3/4)(n^2 - n^3) \ldots]$
$D' = (35an^3/48)[1 - n + (11/16)(n^2 - n^3) \ldots]$
$E' = (315an^4/51)[1 - n \ldots]$

The USGS gives this form, which may be more appealing to some. (They use M where the Army uses S.)

$$M = a[(1 - e^2/4 - 3e^4/64 - 5e^6/256 \ldots)\text{lat}$$
$$- (3e^2/8 + 3e^4/32 + 45e^6/1024 \ldots)\sin(2\text{lat})$$
$$+ (15e^4/256 + 45e^6/1024 + \ldots)\sin(4\text{lat})$$
$$- (35e^6/3072 + \ldots) \sin(6\text{lat}) + \ldots)] \text{ where lat is in radians.}$$

Converting Latitude and Longitude to UTM. All angles are in radians:

$x = \text{easting} = K1 + K2p^2 + K3p^4$, where:

$K1 = Sk_0$,
$K2 = k_0\sin^2 1''$ nu $\sin(\text{lat})\cos(\text{lat})/2$
$K3 = [k_0\sin^4 1''$ nu $\sin(\text{lat})\cos^3(\text{lat})/24][(5 - \tan^2(\text{lat}) + 9e'^2\cos^2(\text{lat})$
    $+ 4e'^4\cos^4(\text{lat})]$

$y = \text{northing} = K4p + K5p^3$, where:

$K4 = k_0\sin 1''$ nu $\cos(\text{lat})$
$K5 = (k_0\sin^3 1''$ nu $\cos^3(\text{lat})/6)[1 - \tan^2(\text{lat}) + e'^2\cos^2(\text{lat})]$.

Easting x is relative to the central meridian. For conventional UTM easting, add 500,000 metres to x.

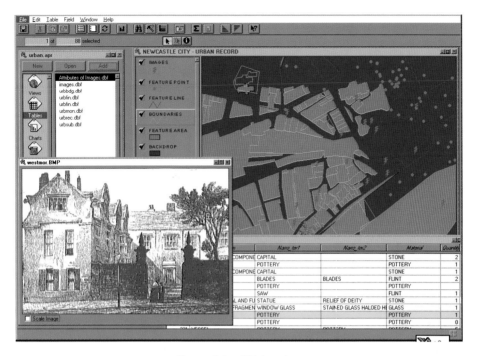

Figure 1.1   GIS interface

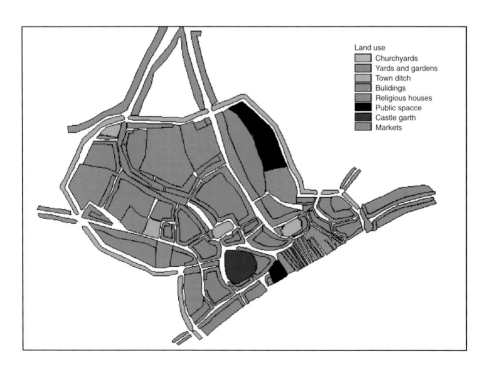

Land use
- Churchyards
- Yards and gardens
- Town ditch
- Bulidings
- Religious houses
- Public spacce
- Castle garth
- Markets

Figure 1.3   Vector land use map

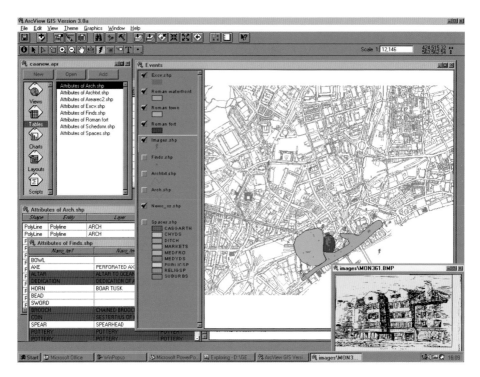

Figure 1.11   *Archaeological GIS*

Figure 4.5 *Control software*
Reproduced with permission from the Ordnance Survey

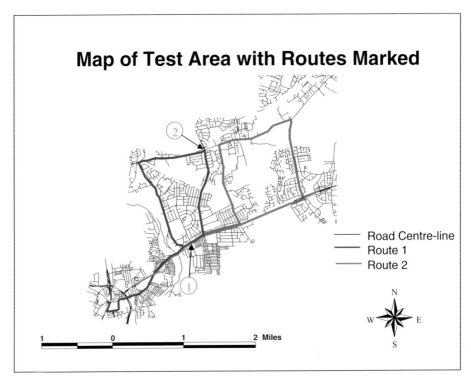

Figure 6.7   *Map of the test area of Newcastle showing the two routes that were used Reproduced with permission from the Ordnance Survey*

3.4.   Orthometric Height 'above Sea Level'

Historically three-dimensional space on the Earth's surface has actually been represented in two dimensions (longitude and latitude), plus a third dimension (height above sea level). This is not a truly integrated three-dimensional system, such as WGS-84. However, such a '2 + 1' system does have perceived advantages, in that the shoreline is (in an average sense) close to zero height, and that it corresponds to the intuitive notion that water flows from higher heights to lower heights.

The disadvantage, historically, is that completely different techniques have been used to survey height benchmarks (typically by spirit levelling) than horizontal benchmarks (typically by triangulation), so two quite separate national networks have typically developed in each country. Moreover, each country generally defines zero height differently. Typically, a benchmark on a specific tide-gauge in each country is used to define the 'height datum'.

In the United Kingdom, for example, the tide-gauge at Newlyn defines zero height. However, across the English Channel in France, it is known that sea level is physically different due to the effect of currents flowing through the Channel. This difference in height definition between the two countries became of practical importance when building the Channel Tunnel. GPS of course was used to solve the problem of tying together the two countries' height systems, and the proof was that the two halves of the tunnel actually did meet in the middle.

GPS and GNSS positions a user in three-dimensional space, and WGS-84 is a true three-dimensional reference system. One way of gaining access to the benefits of 'height above sea-level' is to define everywhere, as part of the reference system, the difference between sea-level and the ellipsoid surface. Technically the the height of the 'geoid' (sea level) would need to be mapped everywhere. There are many ways to do this. One broadly general philosophical approach that seems to be broadly accepted is to base such a map on a 'geoid model', the parameters of which have been tuned to agree with actual measurements at specific locations. That is, such an adopted geoid model would reaonably well explain observed differences in ellipsoidal height (using GPS) and height above sea level at benchmarks of the national height system. The advantage of using a model is that it allows users with a GPS receiver to convert ellipsoidal height into height above sea level no matter where they are.

Since geoid models of this nature will differ from system to system, due in part to real undulations of the Earth's gravity field, but also due to systematic error in historical levelling measurements, it is better to call such heights 'orthometric heights' to distinguish them from true height above sea level. Orthometric height is defined very simply according to the above described procedure:

$$H = h + N(\phi, \lambda)$$

where h is the user's ellipsoidal height as determined by GPS, as transformed into the regional ellipsoidal system; H is orthometric height, which is designed to closely approximate heights of the regional (national) height system, which in turn were intended to represent height above sea level; and $N(\phi, \lambda)$ is based on a physical model of the geoid that closely fits measured differences between H and h.

4. Conclusion

Having read this chapter, you now should have an outline understanding of the techniques and methods used to represent the Earth's shape, and the heights of features on it. You should also be familiar with the mathematical relationships between the ellipsoids used to represent the Earth's three-dimensional shape and the two-dimensional maps used to display the features on the Earth, in particular, the procedures for transforming positional information from GPS coordinates to national grid coordinates.

4

Commercial Applications that Integrate GIS and GPS

This chapter examines current commercial applications that integrate GIS and GPS. Many of existing applications are examples that loosely integrate GIS/GPS, such as low cost in-car navigation systems. However, other current industrial applications use embedded or deeply integrated GIS and GPS. This chapter describes applications, other than navigation, for which these combined technologies are now used, for example, Location Based Services (LBS), and road-tolling systems.

1. Introduction

There are now many applications that use GPS observations for accurate positioning, often simultaneously with other observations (e.g. geophysical, agricultural, hydrological, epidemiological), and then use GIS for analysis, sometimes for complex spatial statistical analysis. Deformation monitoring and plate tectonic motion are prime examples where very high accuracy GPS positioning is used to detect movement. Thurston, *et al.* [2003] provide an extensive overview of geotechnology integration including GPS and GIS together with other sensor systems, such as remote sensing and digital photogrammetry. Other widely used applications that loosely integrate these technologies are in-car navigation, systems/intelligent transport Systems (ITS), pedestrian navigation systems, Location-Based Services (LBS), GPS-enabled mobile GIS services, and precision farming. These types of application are discussed in this chapter.

Various organizations are now formally addressing GIS and GPS integration. For example, in the USA, the Cooperative Research, Education and Extension Service (CSREES), together with industrial and educational partners, have

Intelligent Positioning: GIS-GPS Unification G. Taylor and G. Blewitt
© 2006 John Wiley & Sons, Ltd

formed the GIS/GPS Integration Team. This team is addressing issues such as map accuracy, spatial data sources and methodologies for 'Public Information Systems' that integrate GPS and GIS [Information Technology Education, 2005]. In the UK, the Applied Technology Institute (ATI) offers a course on the integration of GIS and GPS. ATI offers a course on Geomatics (GIS, GPS, and Remote Sensing) that provides a technical overview of the current state of geographic information systems and the integration of this powerful tool with GPS, remote sensing, and other data [ATI, 2005]. Various GIS and GPS vendors have also taken the integration of the two technologies seriously. Trimble, a major GPS vendor, provides hand-held GPS Mapping Systems with embedded GIS functionality for GIS data collection and maintenance [Trimble, 2005], Environmental Systems Research Institute (ESRI), one of the largest GIS companies in the world, have published the book *Integrating GIS and the Global Positioning System* [Steede-Terry, 2000], which describes an assortment of integrated applications, including automatic vehicle location, precision agriculture and elephant tracking. Autodesk have recently been hosting a discussion group on integration of GPS with its own mapping/GIS software [Autodesk, 2003].

2.   National GIS/GPS Integration Team

Cooperative Research, Education and Extension Service (CSREES) and a coalition of partners, including the Farm Service Agency, the USDA Forest Service, the USDA Natural Resources Conservation Service, public and private sector software and hardware companies, the American Association of Geographers, and the Environmental Systems Research Institute (ESRI), have formed the National GIS/GPS Integration Team.

The team works to accelerate the adoption of innovative and creative uses of GIS and GPS in making social, economic, and environmental improvements in our communities, states and nation. This nationwide team is made up of university faculty, county government staff, federal government officials, students, farmers, teachers, and elected officials [Information Technology Education, 2005].

3.   GIS and GPS Deformation Monitoring

Continuously operating networks of high precision GPS receivers are being used in many parts of the world to monitor ground deformation due to earthquakes and other activities. GIS is then used to map, display and investigate the observations. Ge *et al.* [2002] describe the integration of GPS, radar interferometry and GIS for ground deformation monitoring. In this paper [Ge *et al.*, 2002] consider that continuously operating networks of GPS receivers (CGPS) may still not be dense enough to monitor certain phenomena, e.g. volcano and ground subsidence due to mining. Therefore the authors propose to combine GPS with radar interferometry (InSAR) and GIS so that CGPS can monitor small-scale deformation:

> The methodology is to use CGPS to estimate the differential tropospheric delays and apply these estimations as corrections to the radar interferometric results in order to ensure sub-centimeter accuracy. The corrected InSAR results are exported to the GIS so that the ground deformation can be interpreted along with other spatial information such as aerial photos and mine plans.
>
> (Ge *et al.*, 2002)

Bell *et al.* [2002] use a combination of GPS, InSAR, and GIS to assess land subsidence in the Las Vegas Valley due to groundwater withdrawal. The GIS is used to infer how ground deformation correlates with known geological faults that penetrate through the aquifer system, and thus assist discovery as to how such faults control hydrological flow.

Blewitt *et al.* [2003] use a combination of GPS and GIS to seek correlations between GPS-inferred regional deformation in the Great Basin (which encompasses the entire State of Nevada and across the border into neighbouring states) with mapped geological faults to infer through a GIS directional analysis that the rate of extension orthogonal to such faults controls regional-scale permeability and hence the location of economically viable hydrothermal systems.

4. Location-Based Services

Location-Based Services (LBS) provide personalized services to the subscriber based on their current position. These services are usually based on cellular telephone networks. There are various methods used for mobile telephone location, including of course GPS. A discussion of various methods of positioning mobile telephones is given by Mountain and Raper [2001], and Pietilä and Williams [2002].

The potential applications for LBS technologies are numerous and, according to Southern [2002], include:

- navigation – routing a user between two points, and tracking them as they follow the route;
- points of interest – locating restaurants, pubs, cinemas and other entertainment venues in the vicinity of the user;
- advertising – sending details of special offers which the user may be interested in when they are in the neighborhood of a particular retail outlet;
- real-time information – traffic and weather notification for the user's current and predicted location;
- workforce management – ensuring that the most efficient use is made of a number of teams of service engineers by tracking their current location and incorporating the results into a logistics application to minimize the distance travelled between call-outs;

Figure 4.1   *Standard (left), and ruggedized (right) hardware for accessing PARAMOUNT services*
Source: Löhnert, et al. [2003]

- asset management – particularly the capture and validation of an asset's location and condition on the ground;
- locating people – family members or friends, for security reasons or for other functionality such as gaming;
- emergency applications – allowing easy location of persons requiring assistance.

In Europe, the 'Public Safety & Commercial Info Mobility Applications & Services in the Mountains' (PARAMOUNT) is a prototype LBS developed for the use of tourists in the Alps and the Pyrenees [Löhnert *et al.*, 2003]. What is more, PARAMOUNT is also a support tool for controlling and coordinating search-and-rescue (SAR) organizations. This is an archetypical GIS-based LBS, in that it has been developed using GPS for location, dedicated to a specific type of end user with tightly focused functionality and is delivered through a mobile Internet connection. Furthermore, it requires only the basic hardware required for almost all LBS, shown in Figure 4.1. Three core services are offered to the tourist: INFOTOUR, SAFETOUR and DATATOUR.

INFOTOUR is for guiding, routing and informing the user. It provides basic services, which allow the user to access dedicated information at home and in particular on the mountains. SAFETOUR provides the additional facility for a mountaineer to send an emergency call including their current position in case of emergency. SAFETOUR can also track a mountaineer in critical environments and in dangerous terrain. DATATOUR directly involves registered users

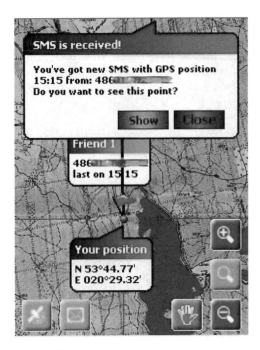

Figure 4.2 *Find a Friend GUI*
Source: Matt [2004]

in the acquisition and maintenance of trail and points of interest data held in the PARAMOUNT system.

The book *Location-Based Services* [Schiller and Voisard, 2004] contains a chapter dedicated to LBS applications and includes a case study on the 'Find a Friend' mobile phone application. 'Find a Friend' applications are provided by a number of cellular telephone networks, they exchange the position point provided by GPS using the SMS cellular phone network. A user can locate a friend and see their position on a map or satellites photo, thus determining how far they are away from the user's current position [Matt, 2004]. A typical Find a Friend GUI is shown in Figure 4.2.

The Location-Based Services Resource Center [LBS] provides links to an abundance of information on Location-Based Services (LBS), E911 and Location-Enabled Business (LEB), including many applications.

Udani and Goel [2002] describe a GPS-enabled mobile GIS service for real-time data collection. This paper discusses requirements for various GPS-enabled mobile GIS application areas, including:

- real-time data collection for dynamic applications related to disaster;
- mapping of infrastructure facilities like school, post office, health centre, bank, veterinary dispensary, bus stop, police station, electric power pole, etc.;

- visualization of villages and population affected based on flood condition;
- visualization of relief and rehabilitation centres and routes for approach;
- position of drinking water sources and their service zone;
- handy tool for health worker who has to go door to door for data collection;
- mapping of forest boundary, area affected by forest fire, area under energy plantation;
- military fieldwork;
- provide information in tabular form regarding agencies and individuals with their resources and responsibilities for the purpose of the disaster management;
- road network mapping within village and connectivity to relief centre;
- user friendly forms for socio-economic data collection for GIS databases.

Navigation systems for pedestrians have requirements that are very different from those for vehicles. A vehicle will in most circumstances be in the open with a reasonable sky visibility; this is not true for a pedestrian. Such navigation systems must integrate not only intelligent maps (GIS) and GPS, but other positioning technologies, such as a dead reckoning device, perhaps a digital magnetic compass, accelerometer and altimeter. GIS can provide map matching and Kalman filtered (navigation filter) trajectory tracking to augment position determination. Gartner *et al.* [2005] describe such an approach based on a proprietary hand-held personal computer, the results compare pedestrian positioning using raw GPS, raw dead reckoning and a combined solution. A mobile multi-modal interaction (M3I) platform, also based on a Pocket PC combines mobile 2D- and 3D-graphics with synthesized speech generation, and the fusion of both speech and gesture input through the use of different recognizers [Wasinger *et al.*, 2004].

In 'Three Dimensional Fuzzy Logic Based-Map Matching Algorithm for Location Based Service Applications in Urban Canyons', Syed and Cannon [2004], make the observation that ITS (see Section 5), is in fact just another LBS application. Syed and Cannon describe an embedded GIS and GPS three-dimensional map-matching solution for use in LBS applications. This innovative solution makes use of a high sensitivity GPS (HS GPS) receiver, digital mapping and a Digital Elevation Model (DEM). This solution attempts to addresses a problem encountered with HS GPS, when it is used in urban canyons. That is, while HS GPS increases the availability of GPS by tracking weak signals, it often introduces large errors and noise in measurements often due to multipath signals arising from reflections off buildings. Map-aided Receiver Autonomous Integrity Monitoring (RAIM), based on road segment information, is used to exclude pseudoranges that have excessive multipath errors from the least squares estimation of GPS position, see Chapter 2, Section 3. There are various RAIM algorithms, which are contained in the GPS receiver and are used to identify faulty GPS measurements [Kaplan, 1996; Brown, 1996].

5.   Intelligent Transport Systems

ITS are those applications related to vehicle tracking and routing that often allow an organizer to identify the locations of vehicles, and enable the system to respond or instruct as appropriate to the needs of the user and/or controller. These applications range from private car navigation systems, commercial fleet tracking, police cars, ambulances, fire engines, etc., in real time. This is a rapidly growing market worldwide, for example the number of vehicle navigation systems shipped, in Japan, in 2001 was 2,180,000 [Anai and Ikisu, 2002], see Table 4.1 (units are hundred million Yen). The information a tracking system provides is especially important when quick-response decisions must be made based on current vehicle locations. Such a system's capability and usefulness are enhanced significantly when location information is tied to geographic information through a GIS [Chou *et al.*, 2005].

The two essential types of information required for ITS applications are the determination of the vehicle position and the determination of the physical location of the vehicle on the road network. The most common devices used for vehicle positioning are based on GPS and dead reckoning (DR) devices or Inertial Navigation Systems (INS), see Chapter 8. Map-matching techniques are often employed to aid the determination of the location of the vehicle on the road network, see Chapter 6.

Many ITS implementations that combine GPS and DR use Kalman Filter techniques for the integration of the two sensors. One such implementation [Ochieng *et al.*, 2005] uses an Extended Kalman filter (EKF) algorithm for the integration of GPS and a low-cost DR sensor to provide continuous positioning. Map-matching is then used on the GPS/DR output both to identify the physical location of a vehicle on the road network and to improve positioning capability. This particular map-matching technique is split into two parts: an Initial Matching Process (IMP) to identify on which road segment the vehicle is located, and Subsequent Matching Process (SMP) to fix the vehicle position on the selected road segment. SMP utilizes map data, vehicle speed from the positioning sensors and the perpendicular projection of the GPS/DR fix on to the link. This combined approach can reduce position errors significantly, see Figure 4.3. A similar

Table 4.1   *Expectation of ITS market in Japan*

Classification of market	2000	2005	2010	2015	Total amount
Telemetric services	768	9449	24950	47729	309903
On-board equipment	4452	10182	15068	17417	186705
Infrastructure	3594	6500	7470	8470	106546
Total amount	8814	26131	47488	73616	603154

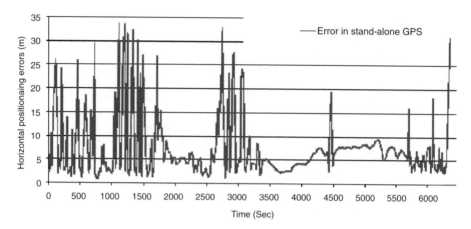

Stand-alone GPS

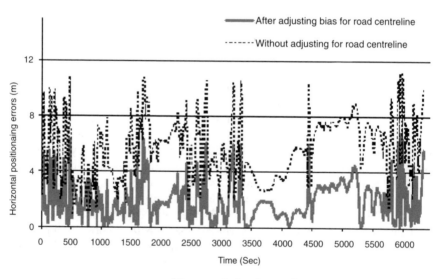

Map-matched results

Figure 4.3 *Horizontal errors of positions relative to the reference (truth) of the vehicle trajectory*
Source: Ochieng et al. [2005]

method of combining GPS and DR is described by Groves and Handley [2004], the difference being that in this implementation terrain-referencing techniques are used to augment GPS and DR, rather than map-matching. Terrain-referenced navigation (TRN) may use a variety of techniques, including terrain contour navigation (TCN) and scene/line feature matching. Groves and Handley discuss various TRN techniques in this paper. The idea is that the addition of one or more TRN systems to an integrated INS/GPS navigation system enables the INS to be calibrated during GPS outages, increasing the robustness of the overall navigation solution.

Mezentsev *et al.* [2002] has researched a vehicular navigation system that consists of an unaided high sensitivity GPS receiver integrated with a navigation quality piezoelectric vibrating gyro. The results of their experiment indicate that new, high-sensitivity GPS receivers provide a position much more frequently in urban canyons (almost 100%, compared to 30%) than standard sensitivity receivers. They also claim that augmentation with a rate gyro significantly improves the 2D position solution accuracy.

6. Accessible Rural Public Transport (Case Study)

6.1. Overview

The current provision of public transport in rural areas is in the main very limited. The form of provision varies from one part of the UK to another, but in general terms there are four main providers:

- private bus companies operating commercial routes;
- the county council public transport co-ordination function which provides conventional routes using a variety of vehicle capacities;
- the county council social services department;
- the voluntary sector/community transport (providing dial-a-ride, rural bus grant, part demand-responsive) with social car, dial-a-ride, minibus and other operations.

Conventional buses are used for primary routes with diversions to off-route villages. Consequently primary route users are discouraged by longer journey times and off-route village residents find one bus per day/week unacceptable and turn to the motor car. A recent report *Bus Travel in Wales: A Consumer's Journey* (Welsh Consumer Council, 2001) showed buses as being a 'last resort' with no 'mechanism' to identify needs.

Several studies [Rajé, 2003; Mackett and Titheridge, 2004; Social Exclusion Unit, 2003; DOT, 2000] have indicated the link between accessibility to public transport and social exclusion. Often this is the consequence of infrequent bus services, with little choice of departure time and long journey times; thus for many it is the 'mode of last resort'. It is desirable to make long distance main route bus services attractive to existing car users.

The departure time choices are a crucial factor in reducing social exclusion. At present, many settlements have one departure per day or even per week as companies and county councils try to provide a basic service. A scheme such as this would provide, through its demand-responsive service, within one-hour travel windows, up to 14 departures per day in each direction between 7 a.m. and 11 p.m.

At present, various county council departmental service providers operate almost independently of one another, often duplicating running over certain routes, e.g. the section that licenses taxis also issues public house licences, but not bus licences. Many transport services in the area are not demand-responsive with public and educational bus services operating on a fixed timetable.

One case study, conducted in Wales, identified an existing timetabled journey time varying from 40 minutes for the direct service to 75 minutes for those serving settlements with small populations located off, but adjacent to, the main trunk route linking two towns, A and B with railway stations, only 20 miles (32 km) apart.

In place of the fixed timetable approach used at present, one solution is to link conventional bus and train trunk route services to an integrated, demand-responsive, low capacity service, for off-route villages. The direct service between towns A and B would operate on an hourly (in place of the present 2-hourly frequency) basis, with a journey time of about 45 minutes but linking into the railway station as well as the commercial centre of both towns. In particular, it would facilitate a far wider range of departure time options for those off the main route travelling via local hubs, providing seamless interchange (including park and ride) facilities. This increased frequency would increase travel options for non-car owners. For one small village, the potential departure times per week (PW) would rise from 4PW to 72PW to town A or B, based on a 12 bus journeys per day.

6.2.   Integrated Rural Transport

The use of GPS and GIS in rural public transport provides the technology to underpin a demand-responsive solution to improve public transport accessibility in rural areas. Passengers can change modes, from low capacity on-demand buses to trunk route buses, at local hubs, where good quality travel information will be available. Synchronized multi-modal travel and on-demand travel information will be achieved using GPS for real-time vehicle tracking and positioning of each vehicle in the system. This vehicle position information is handled at a control centre, where it is used to provide information to drivers, so that interchange from one mode of travel to another may be harmonized. That is, the control centre will automatically inform buses/drivers of the whereabouts of other vehicles in the system. Furthermore, because the position of each vehicle in the system is always known by the control centre, passengers already in the system and waiting passengers may obtain up-to-date travel information by telephone at all times.

6.3.   Route Tracking System

The system consists of a GIS control base integrated into a cellular telephone network. The control GIS requests and receives the position of each vehicle at

Figure 4.4   *System hardware*

set time intervals using mobile phone Short Messaging Service (SMS). It then displays these positions in a GIS. Bus positions in areas in the test region, where GPS availability is problematic, are resolved using Intelligent Map Matching, see Chapter 6. The GIS is programmed to alert the controller of any failure/delay in the service and also stores historical tracking data. Passengers will book a ride, on demand, by phone, either for a time in advance or ad hoc on the day of travel. UK postcode spatial data is used to help identify user pick-up and drop-off points. On-demand vehicles will have their route time optimized, utilizing road centre-line data, to pick up passengers, and synchronize with trunk route buses. A route will be optimized on-the-fly to accommodate last-minute requests. In the UK, mobile phone outages are relatively rare and of short duration, therefore their impact on the system is not important.

The system maintains appropriate Ordnance Survey large-scale digital map and road centre-line data as a backdrop for vehicle tracking and location. The simple hardware required for this system is shown in Figure 4.4 (readily available in-vehicle tracking units), a laptop or PC to run the control software and GSM phone. Figure 4.5 (Plate 4) shows the user interface of the control software written for the pilot project. The actual display is Ordnance Survey 1:50,000 scale digital mapping of the test area.

6.4. Conclusion

The case study addresses three major objectives of integrated transport: improved accessibility and social inclusion, the operational integration of rural bus and rail services, and the financial integration of passenger transport budgets.

The conclusions which might be drawn from the case study are:

- the existing networks are not integrated to maximize travel opportunities or to provide the best use of capacity;
- the provision of public transport services will improve through the integration of bus/taxi/car operations. There will be higher frequency of operation on the trunk routes leading to increased demand (and revenue) and a more respon-

Figure 4.5 *Control software*
Reproduced with permission from the Ordnance Survey

sive and frequent on-demand service with vehicles dedicated to certain areas feeding into local hubs.

- accessibility is extended to sectors of society without personal transport provision;
- by providing real-time information, to both passengers and vehicle operators, such a scheme will remove some of the physical, sociological and psychological barriers to rural multi-modal travel;
- this form of mode integration, with real-time route optimization, can be extended to other regions of the UK and there are possibilities to feed this research into a wider European context.

Research into the concept, solution and policy outlined in this case study was jointly funded by the National Assembly, Carmarthenshire County Council, First Cymru Buses and the University of Glamorgan.

7.   Real-time Passenger Information and Bus Priority System

Another very good example of the application of integrated GIS and GPS for public transport is in the city of Cardiff, UK (with a similar systems operating in other cities around the world, such as Boulder, Colorado, USA). Every bus on certain Cardiff bus routes is equipped with GPS receivers, which use differential corrections supplied via radio transmissions from the control centre (DGPS) to locate each bus. This combination enables the position of the bus to be updated every 20 seconds over a licensed VHF radio channel. This information is the base from which real-time passenger information, bus priority and bus fleet management are realized.

This location information is used to inform passengers, waiting at bus stops, of the expected time of arrival of a bus on such routes. The system uses electronic signs located in bus shelters to provide real-time passenger information regarding predicted waiting times for services. Bus position information is also used to control the operation of traffic lights in Cardiff City centre at 46 signalized junctions [Cardiff City Council, 1999]. An extension to real-time waiting time display at bus stops is to supply this information over cellular phone networks, for specific bus routes and bus stops. If the position of the phone was known, this information could be appropriate to a caller's position and be a true LBS.

8.   Precision Farming

The overall aim of precision farming is to collect spatially referenced data using GPS, GIS and remote sensing techniques, perform spatial analysis and assist decision-making and apply variable rate treatment [Wilson, 1999]. Modern precision farming is implemented using farm vehicles, such as tractors and combine harvesters, with on-board computers interfacing with a satellite navigation system (GPS) to pinpoint precise coordinates within a field. In a simple application, precise coordinates can be used with a yield monitor on a combine harvester to measure within-field yield variability. 'A database of yield variability over time can be used with other information, such as a topographical map of the field, to make crop management decisions for the specified field' [English *et al.*, 2000].

More sophisticated precision farming applications use soil sampling at precise coordinates to determine water content, nutrient content, electro-conductivity, etc. These attributes are stored in a GIS database. This data can then be used, again with GPS positioning for variable rate application of crop-related inputs such as seed, lime, fertilizer, herbicide and pesticide. Precision farming is comprehensively described by Usery *et al.* [1995] and Garten [2003].

GIS for precision farming requires careful design, and software developers must meet a number of challenges: farmers may not have skills in spatial data

Farmer/Consultant

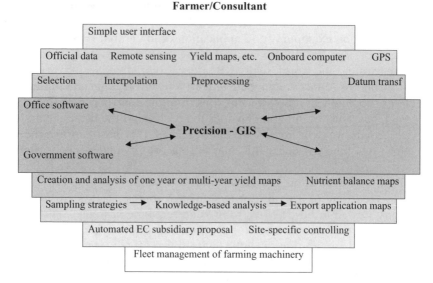

Figure 4.6   *Special requirements for a precision farming GIS*
Source: Grenzdorffer [2000, p. 14]

handling; digital information such as soil maps may not be available; hardware components, including job controllers and on-board computers in the tractor cab have already been developed in the industry and must be supported by the GIS [Grenzdorffer, 2000]. This has led to the development of specialized applications that perform spatial analysis. Figure 4.6 summarizes the special requirements for a precision farming GIS.

Precision farming is now a widely used GIS/GPS technology, with various commercial systems available. There are, of course, many other potential and current applications for integrated GPS and GIS distribution of resources, such as local authority winter maintenance systems (salt and grit spreading on highways), forest fire command and control and battlefield SDSS.

9.   Conclusion

There are a plethora of applications and a few of the more common ones have been described. These applications that integrate GPS and GIS technologies can be coarsely divided into two groups: shallow and deep integration.

9.1.   Shallow Integration

In the implementation of these technologies, GPS receiver positioning is used in GIS to display features, attributes and perhaps simple spatial analysis, such as non-real-time shortest path analysis. Shallow integration can also be used for the following:

- point and other topographic feature capture;
- post-processed GPS imported into GIS;
- real-time field update (GIS background mapping);
- vehicle tracking;
- bus arrival times;
- personal security;
- deformation monitoring;
- precision farming;
- LBS;
- road-tolling.

9.2. Deep Integration

In this combination of these technologies, GPS and GIS data are integrated, perhaps with data from other sensors such as DR instruments to improve positioning accuracy and reliability, and/or combined to provide value-added information for modelling and prediction. Deep integration offers the following features:

- real-time route optimization;
- satellite availability prediction;
- enhanced GPS availability;
- multipath signal modelling;
- map matching algorithms.

The development of integrated applications will inevitably continue to grow, and become more widely used, especially when the European Galileo GNSS is fully operational. The European satellite navigation system Galileo is expected to be in full operation by 2010. It will represent a significant addition and improvement to GPS for all applications and systems that currently integrate GPS and GIS.

5

GPS-GIS Map Matching: Combined Positioning Solution

This chapter describes a novel method of map matching using the Global Position-ing System (GPS), which uses digital mapping data to infer systematic position errors in the raw GPS positions, which can potentially be as large as 100 m if "selective avail-ability" (S/A) were to be imposed by the US military. The system tracks a vehicle on all possible roads (road centre-lines) in a computed error region, then uses a method of rapidly detecting inappropriate road centre-lines from the set of all those possible. This is called the Road Reduction Filter (RRF) algorithm. Point positioning is com-puted using pseudorange measurements direct from a GPS receiver. The least squares estimation is performed in the software developed for the experiment described in this chapter. Virtual differential GPS (VDGPS) corrections are computed and used from a vehicle's previous positions, thus providing an autonomous alternative to DGPS for in-car navigation and fleet management. Height aiding is used to augment the solution and reduce the number of satellites required for a position solution. Ordnance Survey (OS) digital map data was used for the experiment in the UK, i.e. OSCAR 1-m resolution road centre-line geometry and Land Form PANORAMA 1:50,000, 50m-grid digital terrain model (DTM).

Testing of the algorithm is reported and results are analysed. Vehicle positions pro-vided by RRF are compared with the "true" position determined using high preci-sion (cm) GPS carrier phase techniques. It is shown that height aiding using a DTM and the RRF significantly improve the accuracy of position provided by inexpensive single frequency GPS receivers.

1. Introduction

The accurate location of a vehicle on a highway network model is basic to any in-car-navigation system, personal navigation assistant, fleet manage-ment system, National Mayday System [Carstensen, 1998] and many other

Intelligent Positioning: GIS-GPS Unification G. Taylor and G. Blewitt
© 2006 John Wiley & Sons, Ltd

applications that provide a current vehicle location, a digital map and perhaps directions or route guidance. A great many of these systems use GPS to initially determine the position of a vehicle.

As we have seen, GPS has become the most extensively used positioning and navigation tool in the world. GPS provides civilian users with an instant (real-time) absolute horizontal positional accuracy of approximately 10 metres, however, as described in Chapter 2, the US Department of Defense reserves the right to switch on S/A, an intentional dithering of the GPS timing signal. Under such circumstances errors at the level of 100 metres can be expected (corresponding to the level of accuracy when S/A was last switched on). This level of positional accuracy is insufficient to ensure that a vehicle's location will correspond to the digitally mapped road on which the vehicle is travelling. As noted in Chapter 2, developing navigational algorithms to deal with the potential S/A problem will also mitigate unintentional problems with satellite clock behaviour.

A number of methods have been successfully developed to significantly improve GPS accuracy, the most notable being differential GPS (DGPS). In DGPS two receivers work together, one knows its exact position, monitors the errors in its position as provided by GPS, and transmits corrections for these errors to the other receiver. Real-time DGPS can improve positional accuracy down to 1 to 5 m. However, the use of real-time DGPS in a moving vehicle requires additional data in the form of pseudorange corrections, i.e. computed errors in the satellite range measurements. Continuous reception of terrestrial radio transmissions or communication satellite broadcast is required to receive these corrections.

Often data can be combined from multiple sources integrating GPS with other navigational tools, attitude sensors such as the gyrocompass, vehicle odometer, flux gate compass and other dead reckoning methods. This use of multiple data sources again helps to correct for the error (noise) on the GPS position output. Multiple sensor data integration algorithms for vehicles are discussed by Mattos [1993]. Dead reckoning produces the observed track by adding together the position vectors received from the sensor processor [Collier, 1990].

The fact that vehicles are generally constrained to a finite network of roads provides computer algorithms with digital information that can be used to correlate the computed vehicle location with the road network. This is known as map-matching. Many methods have been devised for map matching [Scott, 1994; Mallet and Aubry, 1995]. Our research has developed and tested an algorithm that utilizes GPS for the initial vehicle position and geometric information, computed from the digital road network itself, as the only other source of data for map-matching.

2. Map-Matching Methodologies

Map-matching techniques vary from those using simple point data, integrated with optical gyro and velocity sensors [Kim *et al.*, 1996], to those using complex mathematical techniques such as Kalman filters [Tanaka *et al.*, 1990].

A semi-deterministic map-matching algorithm, described by French [1997], assumes that the vehicle is always on a predefined route or road network. The algorithm determines where the vehicle is along a route or within the network by determining instantaneous direction of travel and cumulative distance. This is a dead reckoning system, driven by interrupts from differential odometer sensors installed on the left and right wheels. The system uses the digital road map to check for correct left or right turns and to remove distance measurement. The positional error is converted into along-track and cross-track errors, allocating the first to the distance sensor and the second to the heading sensor errors [Mattos, 1993]. For example, if the sensors indicate a 90-degree left turn and the digital mapping confirms this with the vehicle's current position, the distance count may be reset to zero. Dead reckoning and map-matching systems like this are often linked with GPS receivers through software filtering schemes such as Kalman filtering [Levy, 1997].

Scott [1994] provides a mathematical framework for map-matching of vehicle positions using GPS. In this framework, the theoretical performance of a map-aided estimation process is assessed using error statistics to translate the raw positions onto the road network. However, Scott acknowledges that a key component of the map-aided estimator is correct road identification. All performance measures derived for the estimator are not applicable if the vehicle position has been projected onto the wrong road. This is true for performance measures of any map-matching algorithm.

Systems that use only geometric information must utilize the "shape" of line segments (road centre-lines) that define the road network [Bernstein and Kornhanser, 1998]. A logical first step is to determine which road centre-lines are candidates for the vehicle's true location. All road centre-lines that cross the region of the possible true position must be located. For example, there are eight potential centre-lines within the 80 m region displayed in Figure 5.1. This region will vary from a 100 m radius circle (when S/A was switched on) with the centre at the computed raw/uncorrected GPS point position (estimated position in Figure 5.1) to a small error ellipse centred on the corrected position, with typical semi-major and semi-minor axis of 5 m and 3 m respectively. The shortest Euclidean distance from the GPS position to each of these road segments is computed and ordered by decreasing distance. The method used here must first calculate the coefficients A, B and C for the implicit and normalized equation of a line through two points that define the road centre-line (line segment):

$$Ax + By + C = 0$$

Let the line be described as going from point k to l. If coordinates l_E, l_N are grid Easting and Northing for point l, and k_E, k_N are Easting and Northing for point k, then:

$$A = k_N - l_N$$
$$B = l_E - k_E$$
$$C = k_E l_N - l_E k_N$$

If A and B are both zero, then it is a bad line definition, otherwise:

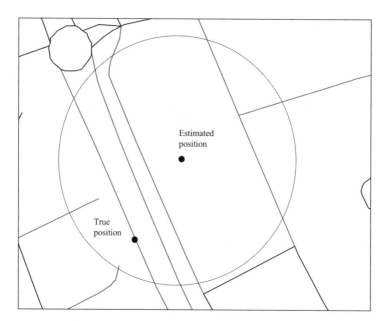

Figure 5.1 *Potential road centre-lines within 80 m of estimated position
Reproduced with permission from the Ordnance Survey*

$$A = A/\sqrt{A^2 + B^2}$$
$$B = B/\sqrt{A^2 + B^2}$$
$$C = C/\sqrt{A^2 + B^2}$$

Then calculate the shortest distance D from the GPS position p to the line that is defined by $Ax + By + C = 0$ using the formula:

$$D = |Ap_E + Bp_N + C|$$

If D is equal to zero, then p is on the line. In addition, if D is positive, p is to the right of the line, or if D is negative, p is to the left of the line joining k to l. This information is of use to map-matching algorithms, and is used here.

It is not simply a matter of finding the line segment nearest to position p. This will often give an incorrect result. For example, in Figure 5.1, the vehicle is on the highlighted road segment in the SW corner, but this is not the nearest road line segment to the position p. Given a single GPS point position for the vehicle, it is not possible to determine which road segment is correct, if there is more than one road segment in the neighbourhood of the possible true position.

A better way to proceed is to match arcs defined by a series of GPS point positions $\{p_1, p_2, p_3 \ldots, p_n\}$ with an arc defined by a set of points that define a partial road centre-line $\{c_1, c_2, c_3 \ldots, c_n\}$. One method used for matching two curves (arcs) is to use the distance between them (White *et al.*, 2000).

If P and C are two such arcs, the minimum distance between any pair of points taken from each arc, i.e. may define the shortest distance between them:

$$\|P - C\|_{min} = \min_{p \in P,\, c \in C} \|p - c\|$$

Bernstein and Kornhanser [1998] describe a curve (arc) matching algorithm, which uses a measure of the average distance between two arcs. This algorithm is implemented for map-matching in Princeton University's Large-Scale Automobile Routing (PULSAR) Project. Arcs must be parameterized by using a function such as $p : [0,1] \to P$, then

$$\|P - C\| = \int_0^1 \|p(t) - c(t)\| dt$$

The problem with this algorithm is that it will only reliably detect the best match for arcs of the same length, which limits its ability to identify the correct arc in certain circumstances, e.g. slow-moving vehicles.

Moments and moment invariants may potentially be used to match arcs. These properties of shapes are used in digital pattern recognition since they are independent of general linear transformations. Singer [1993] describes a method of moment expansion for linear objects (arcs), which may be used for arc comparisons. The moments associated with a line segment li can be written as

$$\mu(r, s) = \int_{li} x^r y^s dl$$

Appropriate moments of two matching arcs would differ only by some small pre-determined tolerance.

Other methods are also used to reduce the number of potential road segments for the correct vehicle position. The topology of the road network may also be used. If the length of the connected route through a network from the present position on one particular road segment to the next position on another potential road segment is outside the possible range of distance travelled so far, that potential road segment is rejected. Carstensen [1998] has looked at the effects of filtering autonomous GPS points by number of satellites tracked, Dilution of Position (DOP) values or satellite geometry, and velocity and acceleration of the vehicle between positions. Filtering of potential road segments may be achieved using some of these measures and other criteria such as distance travelled and change of heading between vehicle positions; both criteria are used here.

3. Road Reduction Filter (RRF) Map-Matching Algorithm

3.1. Introduction

The method of map-matching developed here is dependent on two main innovative techniques. The computation algorithm of a least squares estimation of the position, described in Chapter 2, provides complete control over which

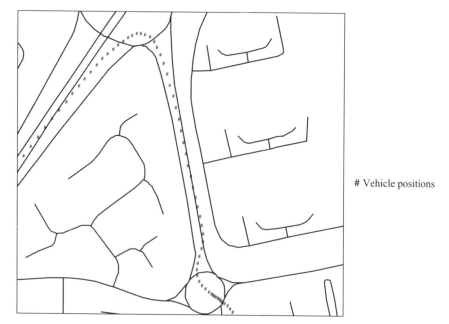

Vehicle positions

Figure 5.2 *Raw GPS positions and road centre-lines*
Reproduced with permission from the Ordnance Survey

satellites will be used in the solution. This avoids step functions in the GPS positions as a result of the loss and gain of satellites. It enables the use of height
aiding in the solution, i.e. removes one of the unknown parameters, thus one less
satellite is required for the computation. The resolution and accuracy of the
DTM must be sufficiently high, or a significant error will be introduced in the
position estimation. Computation of position using pseudoranges also enables
the calculation and use of pseudorange corrections, derived from the digital road
network data virtual DGPS (VDGPS).

A method of modelling Selective Availability (S/A) is described in Chapter 6,
although in the long term S/A-introduced error will reduce to approximately a
Gaussian distribution (random error), in the short term (20 seconds) the effect
of S/A can be viewed as a slowly varying bias [Scott, 1994]. S/A will move the
point position of a stationary receiver by approximately 10 m to 30 m per minute.

3.2. The Algorithm

The general approach adopted in this work to improve the accuracy of the
computed position of a vehicle is to identify all possible candidates for the
correct road and systematically remove the wrong ones. This is achieved by map-
matching; all candidate road segment arcs are compared with the trajectory
described by the Raw (uncorrected) point positions. This is based on the distance
travelled and bearing of the Raw point positions compared with the correspon-

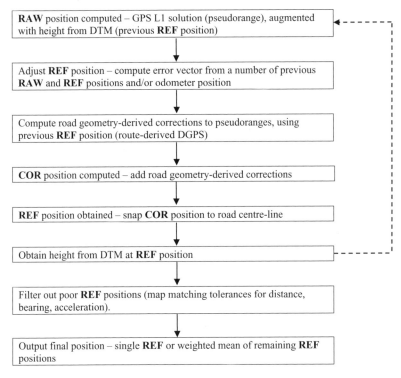

Figure 5.3   *Algorithm flow chart*
Notes: RAW position = GPS derived position
COR position = RAW position corrected with road geometry-derived pseudorange corrections, which are computed using previous adjusted REF position
REF position = COR position snapped to road centre-line

ding Ref positions on road centre-lines. When vehicle velocity is low, this approach is unreliable, and roads are not removed. The algorithm flow is given in Figure 5.3, and is as follows:

1 A Raw vehicle position is computed using all satellites available plus height aiding, where height is obtained from a DTM, which is used to provide an extra equation in the least squares approximation computation, i.e. computation with a minimum of three satellites. For the first epoch all roads (road centre-line segments), which are within 100 m of the computed Raw position are selected. We can guarantee with 95% confidence that the vehicle is on one of these road segments, according to GPS specification [Department of Defense, 1992]. The point on each of the n road segments that computes the shortest distance to the Raw position, using the above equations, is selected as the first approximation of the true location of the vehicle, its Ref position. That is, we have n Ref positions that we can use to generate virtual DGPS corrections for use with the next epoch's computed Raw position.

2 Virtual DGPS corrections for each satellite pseudorange are computed at each of the n Ref positions on each road segment for the current epoch, giving n different sets of virtual DGPS corrections.

3 The next epoch Raw position is computed, as in step 1.

4 Each of the virtual DGPS corrections (step 2) are added to the Raw position (step 3) to give n Cor positions for each n road segments.

5 Each of these n Cor positions is now snapped back onto the nearest road centre-lines to give n Ref positions. Go to step 2.

Steps 2 to 5 are repeated continuously. The output point position from the RRF is either taken from the only remaining road centre-line or is the weighted mean of points on all candidate road centre-lines. The weighted mean position is calculated as follows:

$$\text{Easting} = \frac{\sum \dfrac{E_{REF}}{\text{Average Bearing error}}}{\sum \dfrac{1}{\text{Average Bearing error}}}$$

$$\text{Northing} = \frac{\sum \dfrac{N_{REF}}{\text{Average Bearing error}}}{\sum \dfrac{1}{\text{Average Bearing error}}}$$

where E_{REF} and N_{REF} are Easting and Northing for each of the n Ref positions. These are inversely weighted by the Average Bearing error (described below), which positions the vehicle closer to the road with the smallest bearing error. This helps to smooth point positions when the one correct road centre-line has not yet been found.

When the correct road is found, only that one is processed, until a road junction is encountered, when again as in step 1, a number of road centre-lines that cross the computed error region would have to be considered. The computed error region is an error ellipse calculated from the elements of the cofactor matrix for the estimated parameters in the least squares computation of the position (Chapter 2). The observation error variance used to scale this ellipse is dependent on the Distance error and Bearing error (described below), i.e. the larger these errors, the larger the dimensions of the error ellipse.

This process from steps 2 to 5 is repeated for each new epoch. At each epoch for each of the n road segments the following data is computed and stored:

> Raw distance – from previous Raw to current Raw position.
> Ref distance – from previous Ref to current Ref position.
> Raw bearing – from previous Raw to current Raw position.
> Ref bearing – from previous Ref to current Ref position.

This data is held for the last 20 one second epochs for each road centre-line processed.

Distance errors

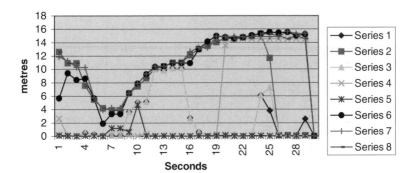

Figure 5.4 *Distance errors for tracked road centre-lines*

3.3. Determining the Correct Road Centre-Line

The task of the RRF is to determine the correct road centre-line segment from the set of those possible, or conversely, which road centre-lines to reject. The trajectory defined by Raw GPS positions computed from observations taken by a receiver in a moving vehicle is correlated with the shape of the digitized road centre-line on which the vehicle is travelling, see Figure 5.2. This correlation is high if the vehicle is travelling at high speed, and low if the vehicle is travelling at low speed (5 mph or 2 m/s), because at high speed any error bias changes less per unit distance travelled.

By calculating values for distance travelled and bearing between epochs for Raw positions and comparing these values with equivalent Ref positions, it is possible to filter out many incorrect road centre-lines. There is a very high correlation if we compare bearing and distance between successive Raw positions and the bearing and distance between successive Ref positions on the correct road centre-line, if the vehicle is moving. We can calculate the following errors for each epoch:

1 Distance error = absolute value of the difference between Raw distance and Ref distance.
2 Bearing error = absolute value of the difference between Raw bearing and Ref bearing.
3 Average bearing error = average of 2 (bearing errors) for the last 20 positions.

Figures 5.4 and 5.5 compare errors for eight different series of Ref positions, i.e., eight different road centre-lines, over a period of 20 epochs (1-second interval). Figure 5.4 displays the distance errors and Figure 5.5 the bearing errors. The bearing errors are clipped at 90 degrees to provide a graph at a scale that is legible. The filter is switched off. If a pre-defined tolerance value is set, perhaps 5 m, it can be seen in Figure 5.4 that series 2, 6, 7 and 8 can quickly be filtered

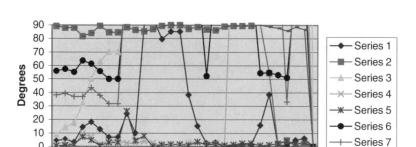

Figure 5.5 *Bearing errors for tracked road centre-lines*

out at almost any epoch. Series 3 and 4 are not as easy to identify for removal, and series 1 is even more difficult to remove. Series 5 is the set of Ref positions on the correct road centre-line, this has a maximum error of 5 m at epoch 10, where the vehicle is driven through a roundabout, see Figure 5.1. Series 3, 4, 1 and 5 are in fact positions from the four parallel road centre-lines displayed in Figure 5.1. Similarly for the bearing errors in Figure 5.5, a filter value of 30 degrees would remove all but the correct series of Ref positions. The Raw, Cor and Ref positions for this correct road centre-line (series 5) are shown in Figure 5.6.

4.   Testing VDGPS

4.1.   Testing Methodology

To evaluate the performance of the algorithm, data was collected in a vehicle driven on roads in the suburbs of Newcastle upon Tyne, UK. An Ashtech Z12 receiver was used in the test vehicle with an Ashtech 700718B Geo.III L1/L2 antenna mounted on the roof. GPS data was recorded at 1-second intervals. Over the same period, a static Ashtech Z12 receiver recorded base-station data on the roof of the Department of Geomatics, University of Newcastle. The data sets were then post processed a number of times, using:

1 Only software developed during this project, using C/A code observations from the vehicle and the OS digital map and DTM data.
2 GPSurvey software using dual frequency phase data from both the vehicle and the base station to compute a high precision (cm accuracy) GPS solution, i.e. Real Time Kinematic (RTK) processing.

Both methodologies have used the following common settings, 15-degree elevation mask, Saatsamoinen tropospheric model and no cut-off value for

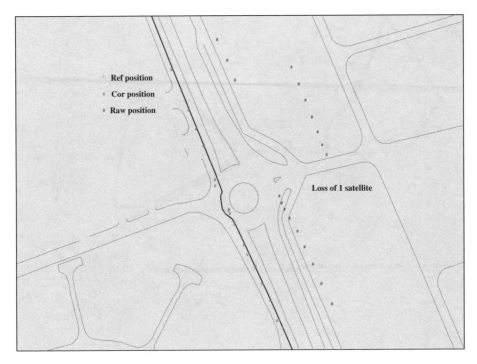

Figure 5.6 *Correct road centre-line only*
Reproduced with permission from the Ordnance Survey

Position Dilution of Precision (PDOP). DOP modifies the ranging error to a satellite, and it is caused solely by the geometry between the user and the set of satellites.

The computed point position outputs from method 1 are: uncorrected GPS receiver positions (RAW), uncorrected GPS receiver positions using OS DTM data for height aiding (RAW+HA) and RRF positions using both OS road centre-line data and OS DTM data for height aiding (VDGPS). To correctly display the vehicle positions on OS large-scale mapping, all resultant latitude, longitude and height coordinates from methods 1 and 2 were transformed from WGS84 (World Geodetic System, 1984) to OSGB36 (Ordnance Survey of Great Britain, 1936) National Grid. This was computed using the Ordnance Survey's OSTN97 with a nominal transformation accuracy of 20 cm with respect to OS primary and secondary triangulation networks, and the European Terrestrial Reference Frame 89 (ETRF89) realization of WGS-84. To use height aiding in the least squares solution the Ordnance Datum Newlyn orthometric heights of the OS DTM are transformed to WGS-84 ellipsoid heights using the OSGM91 geoid model, with a transformation accuracy of 10 cm.

The "true" position of the vehicle at each epoch was assumed to be that given by the RTK solution. Comparisons are made between the following point position outputs:

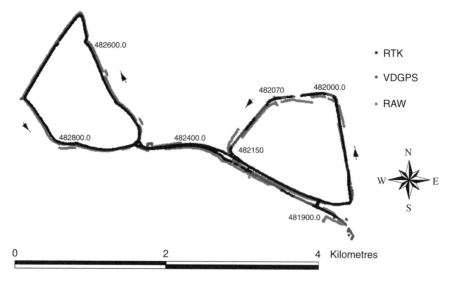

Figure 5.7   *Test route*

Table 5.1   *Processed epochs for each GPS technique*

	RTK	RAW	RAW+HA	VDGPS
processed epochs	1112	1230	1349	1349
percentage	81.6	90.3	99.0	99.0

RTK – VDGPS
RTK – RAW
RTK – RAW + HA

4.2.   Test Results

Figure 5.7 shows the track from each of the three positioning methods. The circuit took approximately 1200 one-second epochs to complete. Table 5.1 summarizes the number of successfully processed epochs (1 second) for each method. All of the position fixes are counted and the total possible number of positions divides the total number of positions recorded, which is the duration of the test in seconds [Bullock *et al.*, 1996]. It can be seen that the use of height aiding, for RAW+HA and VDGPS, provide the best results as far as completeness is concerned, since a solution with one less satellite is possible.

If the RTK positions are assumed to be the truth, horizontal positional errors can be computed as follows

Table 5.2   *Statistics summary*

	VDGPS				RAW	RAW+HA
	Cross-track error	Along-track error	Position error	Position error [95%]	Position error	Position error
Mean	−3.309 m	−0.808 m	14.184 m	11.618 m	35.727 m	26.940 m
Standard Deviation	8.733 m	17.548 m	14.563 m	9.342 m	42.842 m	16.603 m
Minimum magnitude	−65.970 m	−74.856 m	0.018 m	0.018 m	1.345 m	2.367 m
Maximum magnitude	28.464 m	45.259 m	76.272 m	43.543 m	272.541 m	91.898 m

$$DP = \sqrt{DE^2 + DN^2}$$

where DP, DE, DN are the position, Easting and Northing error respectively, and

$$DE = E_{RTK} - E_i$$
$$DN = N_{RTK} - N_i$$

where E_{RTK} and N_{RTK} are the Easting and Northing of the RTK position, and E_i and N_i are the corresponding positions for each of the other methods.

In order to help identify when a wrong road is selected and as an indication of how accurate the algorithm is when the correct road is isolated, the Easting and Northing errors DE and DN of the VDGPS method are transformed into cross-track and along-track errors as follows:

$$\begin{pmatrix} C_i \\ L_i \end{pmatrix} = \begin{pmatrix} \cos t_{i,i+1} & -\sin t_{i,i+1} \\ \sin t_{i,i+1} & \cos t_{i,i+1} \end{pmatrix} \begin{pmatrix} DE_i \\ DN_i \end{pmatrix}$$

where:

C_i is the cross-track error of position i
L_i is the along-track error of position i
DE_i is the Easting error of position i
DN_i is the Northing error of position i
$t_{i,i+1}$ is the bearing of the road segment between position at epoch i and $i + 1$

The mean, standard deviation (SD) and maximum and minimum values are also computed for each error. A summary of the test results, calculated using the previous equations, are shown in Table 5.2. The Position Error (95%) statistics are those of the VDGPS with the largest 5% of Position errors removed (Figure 5.9). The (95%) statistics have much lower values for mean, SD and maximum error, which demonstrates that there are a few very large errors (over 50 m) in the full set of VDGPS results, this is shown in Figure 5.8. Figure 5.11 displays the length of road centre-line on which almost all of these large errors arise, i.e. between epochs 482070 and 482150. The only other time such large errors occur is at the

Position Error (RTK – VDGPS)

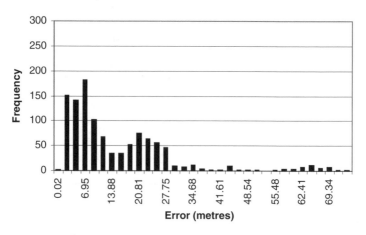

Figure 5.8   *VDGPS position error distribution*

Position Error (RTK – VDGPS) [95%]

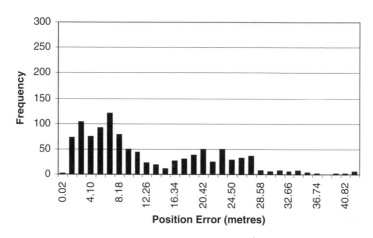

Figure 5.9   *VDGPS position error distribution, 95% confidence interval*

very start of the observation period. In fact, these are both good examples of where the RRF is least effective, when the vehicle is stationary and for a short period after that. That is, before the vehicle moves off at the start of the observations and for approximately 20 seconds at a set of traffic lights. It can be seen from Figures 5.7 and 5.11 that as soon as the road turns, at epoch 482155 approximately, this large position error is rapidly reduced.

Position Error (RTK – RAW)

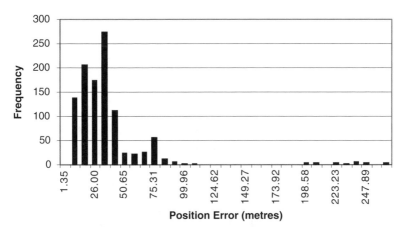

Figure 5.10   *RAW position error distribution*

Position Error (RTK – VDGPS)

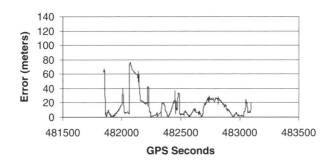

Figure 5.11   *VDGPS position errors*

The main interest here is the comparison of positions provided by VDGPS and those provided by a single GPS receiver with no corrections (RAW). Note that cross-track and along-track errors given in Table 5.2 are those of the full set of VDGPS positions. The worst cross-track errors are again between epochs 482070 and 482150, in fact, at epoch 482090, Figure 5.12, when the vehicle is stationary and the wrong road has been selected by the RRF. The along-track errors, Figure 5.13, are also at a maximum of −80 m near epoch 482090, meaning the car is positioned 80 m behind where it should be on the road. Figure 5.13 also displays how the along-track error gradually decreases as the algorithm-derived VDGPS positions catch up with the true vehicle positions on the road. It is clear

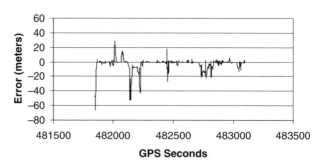

Figure 5.12   *Cross-track errors*

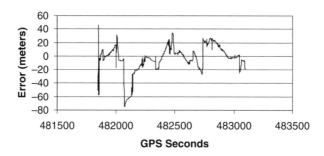

Figure 5.13   *Along-track errors*

from Table 5.2 that VDGPS provides more reduced position errors than RAW, producing an improvement from a mean position error of 36 m down to 14 m, compare the distributions of position errors in Figures 5.8 and 5.10. Both the range of error and SD are also reduced. The very high position errors in RAW (>200 m), shown in Figure 5.10, are a result of not applying any cut-off value for PDOP, this is done to maximize the possible number of epochs for which a position may be computed. If the worst 5% position errors are removed from VDGPS, the results are further improved, see Table 5.2. Figure 5.9 shows the distribution of position errors for VDGPS at 95% confidence interval.

The increasing frequencies of errors between 16 m and 28 m for VDGPS are due to the temporary inability of the RRF to identify the correct road, i.e. for short periods an incorrect road centre-line is being tracked. This could be due to the vehicle travelling at low velocity; the GPS trajectory is unreliable. Cross-track errors are only significantly large again when temporarily the correct road centre-line has not been isolated and a weighted mean position is used.

5.   Conclusions

The augmentation of GPS code data with height and road network information, obtained from digital mapping, for the positioning of a moving vehicle has improved reliability by almost 10%, i.e. 90.3% to 99% of all possible epochs processed. To put it another way, VDGPS reduces the number of positions that cannot be computed from around 10% of all epochs to 1% of all epochs. The mean horizontal position accuracy has changed for the better from 35.727 m to 14.184 m, with good improvements in standard deviation and range of errors, see Table 5.2. If gross errors are removed, these improvements are further enhanced.

This experiment has successfully integrated only satellite navigation and digital mapping data to improve accuracy and reliability of the estimated position of a moving vehicle. The computation combines raw GPS pseudorange measurements, road centre-line geometry and DTM data in new map-matching algorithm. The work has deliberately only used geometry to improve vehicle positioning.

The use of only these two data sources has the distinct advantage of producing a completely self-contained system, requiring no radio communication for differential corrections and continuous data provision, or input from any other sensors. Furthermore, because the computation of the estimated GPS receiver position is part of the RRF, and a digital terrain model-derived height aiding is used in the solution, only three satellites are necessary for a solution.

6

Intelligent Map Matching Using 'Mapping Dilution of Precision' (MDOP)

The exact location of a vehicle on a road is essential for accurate surveying applications. These include close-range photogrammetry using digital video or still cameras and the verification of digital mapping by measured (GPS and other sensors) trajectories. This chapter describes our application of least squares estimation to map matching for vehicle positioning using GPS [Blewitt and Taylor, 2002]. It derives a new method of improving the accuracy and reliability of vehicle positioning provided by GPS by using our patented method to compute an error vector and an associated quality metric called 'Mapping Dilution of Precision' (MDOP). A theoretical proof of the geometric interpretation of MDOP is also given. This method reduces the error in position, which is a sum from several sources, including signal delay due to the ionosphere and atmosphere and until recently from 'selective availability' (S/A). S/A was imposed by the US military to deliberately degrade the accuracy of GPS, but was switched off on the 2nd of May 2000, and is to be replaced with 'regional denial capabilities in lieu of global degradation' [Interagency GPS Executive Board, 2000].

1. Introduction

The identification of the particular road on which a vehicle is travelling may be achieved in a number of ways using map matching and other techniques [Mallet and Aubry, 1995; Scott, 1994; Collier, 1990]. A particular method developed in earlier work, which is built upon here, solves this identification problem using an algorithmic approach. This algorithm, described in Chapter 5, is called a Road Reduction Filter (RRF). This RRF computes certain differences (errors) between the trajectory drawn by raw uncorrected GPS receiver positions taken in a moving vehicle and digital road centre-lines. Potential roads are discarded when distance and bearing differences reach certain tolerances. This method will eventually reduce the set of all potential road centre-lines down to just the correct one, within a few seconds in most cases [Taylor and Blewitt, 2000]. What is less certain with this method is the exact location of the vehicle on that road centre-line. That is, the along-track error will vary considerably.

2.   Least Squares Estimation of Position Error Vector

One problem with the approach briefly described above is that errors in the GPS signal translate into considerable errors in position. It may be possible at a particular point in time to correctly identify the road a vehicle is travelling on, but the position along the road may be in error by up to 20 m (100 m when S/A was switched on). This 'along-track error' cannot be resolved for a straight road, but it can be resolved if the road changes direction, or if the vehicle turns a corner. A more formal method of computing a map-matched correction is now given, which is then integrated with the Road Reduction Filter (RRF). This map-matched correction, or error vector, is used to adjust the position of the vehicle on the road segment, but only when residual values are low. The advantage of formal methods is that quality measures can be derived and used to place confidence bounds for rigorous decision-making (for example, to reject road centre-lines that fail a particular hypothesis test). Formal methods also provide insight into the relative importance of factors, which can improve the procedure (e.g. data rates and road geometry).

Figure 6.1 displays a GPS position at a single epoch. Vector $\underline{b}$ can be considered to be the error vector (position error vector) from the true vehicle position on the road centre-line at grid position $\text{Tru}(E_{\text{Tru}}, N_{\text{Tru}})$ to the uncorrected position computed from GPS at $\text{Raw}(E_{\text{Raw}}, N_{\text{Raw}})$. The perpendicular distance from the Raw position to the road centre-line at $\text{Int}(E_{\text{Int}}, N_{\text{Int}})$ is given as L. The road center-line for this purpose is defined by extending the line segment, which joins the previous Ref to the current Ref. The first approximation of the Tru position is the Ref position, which (as explained in Section 3.2 in Chapter 5: The Algorithm) was obtained by snapping the Cor position (the Raw position corrected using map-matched corrections) onto the closest point on the road centre-line. Furthermore, the observed perpendicular distance from Raw position to the road center-line at Int is given by L where:

$$L = \pm\sqrt{\left(E_{\text{Raw}} - E_{\text{Int}}\right)^2 + \left(N_{\text{Raw}} - N_{\text{Int}}\right)^2}$$

The positive root of L is taken if the raw point lies to the right of the centre-line, and the negative root if it lies to the left. As L has a sign, it may be better described as a 'cross-track coordinate' rather than as a distance.

Here, L is introduced as a 'measurement' which can be modelled geometrically. The model that best fits a series of these measurements provides an esti-

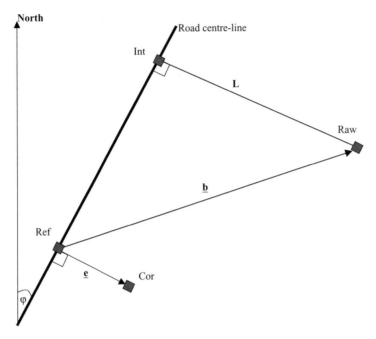

Figure 6.1   *GPS position error vector*

mate of the error vector, $\underline{b}$. Consider the unit vector $\hat{\underline{e}}$ which points normal to the road center-line (and to the right of the road) at the Ref position: the cross-track coordinate L may also be modelled (computed) using the dot product of the two vectors $\underline{b}$ and $\hat{\underline{e}}$:

$$\underline{b} = b_E\underline{E} + b_N\underline{N}$$
$$\hat{\underline{e}} = r_N\underline{E} - r_E\underline{N}$$

where $\underline{E}$ and $\underline{N}$ are unit vectors pointing in the East and North directions, and r_E and r_N are the direction cosines of a road segment at the Ref position. The Ref position is computed using the RRF algorithm. For analytical purposes later, it is convenient to write them in term of ϕ, the bearing (clockwise azimuth from North) of the road segment:

$$r_E = \sin\phi$$
$$r_N = \cos\phi$$

Therefore, an observation equation using the above formula may be formed, where the left side is measured, and the right side is modelled, and includes an unknown term v, which absorbs random position errors:

$$L = \underline{b}\cdot\hat{\underline{e}} + v$$
$$= b_E r_N - b_N r_E + v$$

Such an equation may be formed each time a GPS raw estimated of position is computed. Now consider n successive GPS raw estimates over a time period where the error vector $\underline{b}$ can be assumed to be approximately constant:

$$L_1 = b_E r_{N1} - b_N r_{E1} + v_1$$
$$L_2 = b_E r_{N2} - b_N r_{E2} + v_2$$
$$L_3 = b_E r_{N3} - b_N r_{E3} + v_3$$
$$\vdots$$
$$L_n = b_E r_{Nn} - b_N r_{En} + v_n$$

In practice, $\underline{b}$ varies at a level comparable to a road width over 20 to 50 seconds, dependent on the actual road and vehicle velocity. Hence, for GPS raw estimates every second, n (number of raw position estimates) can have a value of about 30. This can be written in matrix form:

$$
\begin{pmatrix} L_1 \\ L_2 \\ L_3 \\ . \\ . \\ L_n \end{pmatrix}
=
\begin{pmatrix} r_{N1} & -r_{E1} \\ r_{N2} & -r_{E2} \\ r_{N3} & -r_{E3} \\ \vdots & \vdots \\ r_{Nn} & -r_{En} \end{pmatrix}
\begin{pmatrix} b_E \\ b_N \end{pmatrix}
+
\begin{pmatrix} v_1 \\ v_2 \\ v_3 \\ . \\ . \\ v_n \end{pmatrix}
$$

This can be written compactly as:

$$\mathbf{L} = \mathbf{A}\mathbf{x} + \mathbf{v}$$

The principles of least squares analysis is applied, a suitable description is given in Chapter 2, which minimizes the sum of squares of estimated residuals, giving the following solution for (b_E, b_N):

$$\hat{\mathbf{x}} = (\mathbf{A}^T\mathbf{A})^{-1}\mathbf{A}^T\mathbf{L}$$

Note that in above equation, the cofactor matrix $(\mathbf{A}^T\mathbf{A})^{-1}$, also sometimes called the covariance matrix, is implicitly understood to be scaled by the variance of the input observation errors. These errors in this case are characterized by the accuracy of the particular digital road centre-line data used. The focus here is on the cofactor matrix, which like $\mathbf{A}$, is purely a function of direction cosines of road segments, i.e. route geometry.

The estimated residuals (misfit of model to the data) are given by:

$$\hat{\mathbf{v}} = \mathbf{L} - \mathbf{A}\hat{\mathbf{x}}$$

This equation can be used to be checked to assess model fidelity. After the least squares computation it is possible to estimate the precision of the measurements by examining the residuals, i.e. how much the observed values have been altered by the process. If the residual values are low, then this indicates a high precision set of observations [Cross, 1994].

Least squares assume that the errors v_i are random with zero mean expected value (i.e., some will be positive, some negative). It does not depend on the errors being normally distributed. This is a reasonable model for GPS pseudorange measurement error, but is not a good model for persistent systematic effects such as atmospheric delay and errors in satellite positions computed from the Navigation Message. However, such systematic effects will be absorbed by the error vector estimate. Note that such persistent effects are not only common mode to a single receiver's measurements over a short time period, but would also be in common to all GPS stations in the local area. Clearly, the estimated error vector $\hat{x}$ is equivalent to a 'position correction' which could be provided by a local DGPS base station. We call our technique 'map-matched GPS', it does not require data from another GPS base station, but provides the same type of position correction.

Note that the GPS data and the digital map data have been incorporated into this formal scheme through the 'measurement' of L. One advantage of taking such a formal approach to map matching can therefore be seen as the quantification of expected errors, which can in turn be used to narrow down the search for possible positions. For example, alternative hypotheses where a vehicle may have taken one of three roads at a junction can be assessed in terms of the level of estimated residuals, as compared to the level of expected errors.

The modelled error in the determination of the error vector can be found from the covariance matrix $\mathbf{C}$, which can then be used to plot a confidence ellipse within which the true value of error bias can be expected to lie. The covariance matrix is computed as:

$$\mathbf{C} = \sigma^2 (\mathbf{A}^{\mathrm{T}} \mathbf{A})^{-1}$$

The constant σ^2 represents the variance in raw GPS positions, excluding the effects of common mode errors. In other words, σ should equal the standard deviation in raw GPS positions if perfect DGPS corrections were used to remove the effects of nonrandom common mode errors. Its value tends to be dominated by signal multipath around the vehicle, and varies with the geometry of the satellite positions, an effect known as 'horizontal dilution of precision' (HDOP). Typical values are at the metre level. One possibility is to use the estimated residuals themselves to estimate the level of σ. This would be inadvisable, however, because it is intended to use C to test the significance of high levels of residuals, which would have created a circular argument.

3. Quantifying Road Geometry: Mapping Dilution of Precision (MDOP)

The equation given above for the computation of the covariance matrix leads to an elegant method of quantifying road geometry as to its suitability for estimating error in position on-the-fly. First, note that the least squares method assumes that the 'cofactor matrix' $(\mathbf{A}^{\mathrm{T}} \mathbf{A})^{-1}$ exists. It is necessary but not sufficient requirement that $n \geq 2$. If the two **Ref**erence positions are collinear (the road is perfectly straight), then a third position is required that is not collinear. In the work here $n = 30$. We now explore how the cofactor matrix can be interpreted, and how it is related to the shape of the road.

The diagonal elements of the cofactor matrix can each be interpreted as the ratio of the error squared in the estimated error vector component to the expected error squared of a single GPS position if an ideal DGPS position correction were used. To obtain a single number that relates to standard deviation of position (instead of variances and covariances), we follow the example of classic GPS theory by which the square root of the trace of the cofactor matrix is taken as a 'Dilution of Precision' (DOP) value. We therefore define 'Correction Dilution of Precision' (CDOP) as:

$$\text{CDOP} = \sqrt{\text{Tr}(\mathbf{A}^{\mathsf{T}}\mathbf{A})^{-1}}$$

From the previous definition of matrix $\mathbf{A}$, we can write CDOP in terms of the direction cosines at each of the sampled points on the road. Starting with the cofactor matrix:

$$(\mathbf{A}^{\mathsf{T}}\mathbf{A})^{-1} = \begin{pmatrix} \sum_n r_{Ni}^2 & -\sum_n r_{Ei}r_{Ni} \\ -\sum_n r_{Ei}r_{Ni} & \sum_n r_{Ei}^2 \end{pmatrix}^{-1}$$

$$= \frac{\begin{pmatrix} \sum_n r_{Ei}^2 & \sum_n r_{Ei}r_{Ni} \\ \sum_n r_{Ei}r_{Ni} & \sum_n r_{Ni}^2 \end{pmatrix}}{\sum_n r_{Ei}^2 \sum_n r_{Ni}^2 - \left(\sum_n r_{Ei}r_{Ni}\right)^2}$$

Therefore, substitution of this into the previous expression for CDOP gives:

$$\text{CDOP} = \left(\frac{\sum_n r_{Ni}^2 + \sum_n r_{Ei}^2}{\sum_n r_{Ei}^2 \sum_n r_{Ni}^2 - \left(\sum_n r_{Ei}r_{Ni}\right)^2}\right)^{1/2}$$

It can be shown that the numerator is simply n, so the whole formula can be reduced to:

$$\text{CDOP} = n^{-1/2}\left(\overline{r_{Ei}^2 r_{Ni}^2} - \overline{r_{Ei}r_{Ni}}^2\right)^{-1/2}$$

where the overbars denote averaging over the section of road (for which the error is assumed to be approximately constant). CDOP therefore depends on road geometry, and will be inversely proportional to the number of GPS measurements n taken over a fixed time interval. With enough measurements and with sufficient change in road direction, it is possible to reduce CDOP to < 1.

Note that GPS data recording should be sufficient to sample any detail in road shape that is present in the digital map. It is therefore preferable to record GPS data at a high rate, e.g., 1 per second. Going at even higher rates than this would not help particularly, because the road is approximately straight between points. Where there is a detailed road shape, the rate of sampling will increase naturally due to the necessary reductions in vehicle velocity.

A related quality measure is 'Mapping Dilution of Precision', (MDOP) which we define as the ratio of position precision using map-matched GPS to that using perfect DGPS corrections. In this case, we assume that if n is much greater than 1, then the map-matched correction (i.e., the error vector) is uncorrelated with the error in any single data point. Therefore, the corrected position will have a variance equal to the variance in the perfect case plus the variance in the correction. As this is to be divided by the variance in the perfect case, the result is:

$$\text{MDOP} = 1 + \text{CDOP}$$
$$= 1 + \sqrt{\text{Tr}(\mathbf{A}^{\mathsf{T}}\mathbf{A})^{-1}}$$

This measure is particularly useful because:

1 It is easily interpreted as a 'level of degradation' in precision as a result of not using a perfect DGPS base station.
2 It can be tested for validity under controlled conditions.

As we shall describe, testing was carried out using an ultra-precise GPS method (e.g., carrier phase positioning) to determine the true level of corrected position errors, and then compare this with the errors obtained by applying a near-perfect DGPS correction. The point is that the above expression for MDOP can be computed easily in real time (even ahead of time!) by simply knowing the road shape.

Note that MDOP is always greater than 1 because comparison is made with perfect DGPS. It is worth bearing in mind that no DGPS system is perfect; hence MDOP > 1 does not necessarily mean that real DGPS will give better results than map-matched GPS.

4. MDOP for Basic Road Shapes

From the above equations, we can write MDOP analytically in terms of the direction cosines of the vector normal to the road:

$$\text{MDOP} = 1 + n^{-\frac{1}{2}}\left(\overline{\sin^2\phi\cos^2\phi} - \overline{\sin\phi\cos\phi}^2\right)^{-\frac{1}{2}}$$

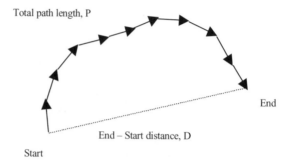

Total path length, P

End

End – Start distance, D

Start

Figure 6.2   *Path constructed of unit vectors*
Reproduced with permission from the Ordnance Survey

This equation can be rearranged into the following form:

$$MDOP = 1 + 2n^{-\frac{1}{2}}\left(1 - \overline{\cos 2\phi}^2 - \overline{\sin 2\phi}^2\right)^{-\frac{1}{2}}$$

The first thing to note about MDOP is that it takes on the following maximum (worst case) and minimum (optimum) values:

$$MDOP_{max} = \infty \qquad \phi = constant$$
$$MDOP_{min} = 1 + 2n^{-\frac{1}{2}} \qquad \overline{\cos 2\phi} = \overline{\sin 2\phi} = 0$$

The maximum condition is satisfied for a straight road. As we shall see, the minimum condition is satisfied for the simple case of a right-angled bend. Bearing in mind the definition of MDOP, we see that GPS error ceases to be a dominant error source when $MDOP \leq 2$, which the above equation satisfies when using four GPS measurements around a right-angled bend. As more measurements are introduced, MDOP approaches 1, which implies that positioning is as good as using a perfect DGPS system.

MDOP can easily be computed for any road using a graphical interpretation of the term we call the 'path closure ratio':

$$S(\vartheta) = \overline{\cos \vartheta}^2 + \overline{\sin \vartheta}^2$$

Consider a path constructed using segments i each of equal length and with bearing ϑ_i (Figure 6.2). The path closure ratio S can be shown to be equal to the square of the ratio of straight-line distance between the starting and end points D to the total path length P:

$$S(\vartheta_i) = (D/P)^2$$

Obviously, S ranges from 0 to 1. We can therefore take our digital map of the road, and transform it to a path where all the path segments have double the bearing of the real road, and where each road segment between GPS points are mapped into segments of equal length. We can then compute MDOP as follows:

$$\text{MDOP} = 1 + 2n^{-\frac{1}{2}}(1 - S(2\phi))^{-\frac{1}{2}}$$
$$= 1 + 2\Big/\sqrt{n\left(1 - (D/P)^2\right)}$$

Note that a path of fixed length P is therefore equivalent to a road section covered in a fixed amount of time (because GPS data are recorded at equal intervals). So for a fixed amount of time, the path which ends closest to the starting point produces a smaller value of S, and a smaller (more favourable) value of MDOP.

This graphical method is so powerful, that results can be visualized without any computation (Figure 6.3). For example, a sharp right-angled bend in a road will map onto a path which doubles back on itself, reducing S to zero, and hence producing the minimum value of MDOP. A road which gently sweeps though 90° will map onto a path which heads back in the opposite direction, but is displaced by some distance, and therefore will produce good, but not optimum results. A road, which goes in a semi-circle (e.g., around a large roundabout), will map into a path, which is a complete circle, and hence will produce optimum results.

Table 6.1 summarizes the results for the computation of the path closure ratio $S(2\phi)$ for various road shapes, which can then be used to find the appropriate MDOP value. Also given is the value of n, which would be required to bring the MDOP value < 2. We call this number the 'resolution time' T, since it tells us how many data intervals are required to bring GPS error to a level below that expected from random position errors. Under the assumption that we use 1 second GPS data, T is in seconds. Alternatively, the value in the final column of

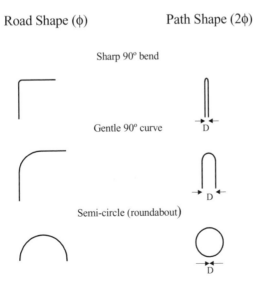

Road Shape (ϕ)   Path Shape (2ϕ)

Sharp 90° bend

Gentle 90° curve   D

Semi-circle (roundabout)

D

Figure 6.3 *Basic road shapes can be transformed into path shapes with twice the curvature, which can then be interpreted in terms of favourable geometry (MDOP)*

Table 6.1   Quality measures associated with road geometries for map-matched GPS

Road shape description	Path closure ratio $S(2\phi)$	Mapping dilution of precision, MDOP	Resolution time T (sec)
Instant bend, angle α	$\cos^2\alpha$	$1+2/\sin\alpha\sqrt{n}$	$4/\sin^2\alpha$
Instant bend, 90°	0	$1+2/\sqrt{n}$	4
Instant bend, 45°	0.5	$1+2.8/\sqrt{n}$	8
Instant bend, 20°	0.88	$1+5.8/\sqrt{n}$	34
Instant bend, 10°	0.97	$1+11.5/\sqrt{n}$	133
Smoothest curve, α	$\sin^2\alpha/\alpha^2$	$1+2/\sqrt{(1-\sin^2\alpha/\alpha^2)n}$	$4/(1-\sin^2\alpha/\alpha^2)$
Smoothest curve, 90°	$4/\pi^2 = 0.41$	$1+2.6/\sqrt{n}$	7
Smoothest curve, 45°	$8/\pi^2 = 0.81$	$1+4.6/\sqrt{n}$	22
Smoothest curve, 20°	0.96	$1+10.0/\sqrt{n}$	100
Smoothest curve, 10°	0.99	$1+19.9/\sqrt{n}$	396

Table 6.1 may be considered to be the minimum number of data points required to describe each road shape, in order to evaluate MDOP.

5.   Testing MDOP

To evaluate the effectiveness of MDOP GPS C/A code observation data was collected in a vehicle driven on roads in the suburbs of Newcastle upon Tyne, UK, see Figure 6.4. Over the same period, dual frequency phase data was collected in the vehicle and also by a static receiver recording base station data on the roof of the Department of Geomatics, University of Newcastle. This dual frequency data was used to compute a high precision (cm accuracy) GPS solution, which was assumed to be the 'true' position of the vehicle at each epoch (second). The details of all hardware, software, data sets and processing techniques are given in Taylor and Blewitt [2000]. All available satellites visible to both receivers were used in the position solution computation (no elevation mask), this number varied throughout the route from none to eight. Three point position solutions were computed:

1 RAW solution   – using C/A code data.
2 Map-matched GPS solution – using C/A code data, the RRF, MDOP and digital map data.
3 RTK solution   – using dual frequency phase data from both the vehicle and the base station to compute a high precision (cm accuracy) GPS solution. The 'true' position of the vehicle at each epoch was assumed to be that given by this solution.

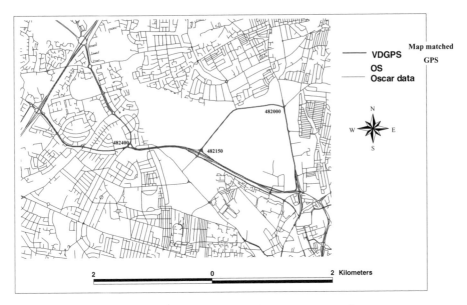

Figure 6.4   *Test route, displaying GPS seconds*

The map-matched GPS positions output from method 2 used Ordnance Survey (OS) road centre-line data, OS DTM data for height aiding RRF for correct road selection and MDOP to correct for along-track errors. To correctly display the vehicle positions on OS large-scale mapping, all resultant latitude, longitude and height coordinates from the three solutions were transformed to OSGB36 (Ordnance Survey of Great Britain, 1936) National Grid, with a nominal transformation accuracy of 20 cm (OS, 1999). At each epoch, where all three positions were available, the difference in position between RAW and RTK and map-matched GPS and RTK were calculated (position error).

A summary of the results is given in Table 6.2. It is also of interest to note that the maximum position errors were 177 m for the RAW data and 76 m for the map-matched GPS data (43 m for 95%). Mean error of position has been reduced from 36 m to 13 m over a total of 1112 vehicle positions. The variation in both cross-track and along-track error is also much reduced. It can be seen that the map-matched GPS described in this chapter provides a much improved accuracy of position, particularly if the worst 5% of position errors are removed. In fact, MDOP can be used to identify (predict) where on the route the error vector will be least accurately modelled. Inspection of the estimated residuals tells us when we have a poor error vector. If the residuals are low, then we can reject a road segment with the RRF.

All the really large errors occur when the vehicle is stationary or almost stationary such as at a road junction, e.g. approximately at GPS seconds 482000, 482400 (both at roundabouts) and at 482150 (motorway slip road). These two

Table 6.2 *Statistics summary of data collected when S/A was switched on*

	Map-matched DGPS				RAW
	Cross-track error	Along-track error	Position error	Position error [95%]	Position error
Mean	−1.0 m	−2.1 m	12.7 m	9.3 m	35.727 m
Standard Deviation	8.733 m	17.548 m	18.9 m	11.3 m	42.842 m

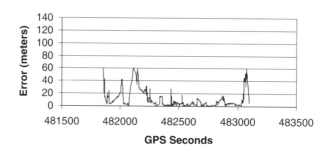

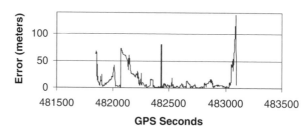

Figure 6.5 *Position errors for map-matched GPS*

positions can be seen on the map in Figure 6.4 and the corresponding errors in Figure 6.5. The only other times are at the beginning and end of the route for the same reason. If we ignore these times when the positions are in gross error, it can be seen in Figure 6.6 that cross-track errors are almost always small, because the vehicle has been positioned by map-matched GPS on the correct road. Along-track errors are larger, as expected, because once a correct road is identified, it takes a number of epochs before the algorithm can successfully use MDOP to correct the position, see Table 6.1.

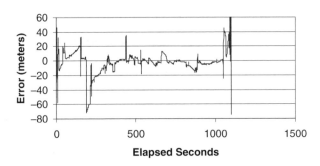

Along Track Error (RTK – Map matched GPS)

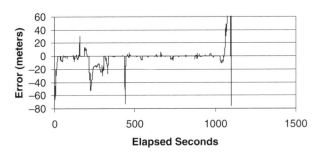

Cross Track Error (RTK – Map matched GPS)

Figure 6.6 *Cross-track and along-track errors for map-matched GPS*

Table 6.3 *Statistics summary of data collected when S/A was switched off*

Error	Map-matched DGPS				RAW
	Cross track	Along track	Position	Pos. [95%]	Position
Mean	−0.28 m	0.22 m	2.40 m	2.02 m	4.58 m
Standard Deviation	1.25 m	2.27 m	2.44 m	1.65 m	2.46 m

A second set of data was collected over the same route, with exactly the same equipment and operational parameters. This data was processed in the same manner as described above. The only significant difference being that the data was collected after S/A had been switched off. A summary of the results is given in Table 6.3. Again, it is interesting to note the maximum position errors were 36 m for the RAW data and 21 m for the map-matched GPS data (7 m for 95%).

6.   RRF Map-matching Enhancement

The RRF map-matching algorithm has been further improved by the addition of a road network tracing routine and the use of intelligent road network data, which contains drive restriction information. A comparison between the existing RRF vehicle tracking algorithm performance and the algorithms performance enhanced with new routines is now given.

The improvement in map matching is achieved by the addition of two subroutines to the existing algorithm. A route network analysis routine, provided with ArcView's Network Analyst Extension, was incorporated into the existing software. This analysis software calculated the distance through the road network from a vehicle's previous position to each potential present position, offered by the map-matching algorithm. Potential positions requiring an impossible network travel distance are instantly rejected using the newly incorporated network analysis software. This improved the reliability of the route tracking system at locations such as roundabouts or dual carriageways, and immediately after the loss of GPS point positions.

A further algorithm was developed so that the RRF could interrogate Ordnance Survey (OS) OSCAR Traffic Manager with Drive Restriction Information (DRI), DRI is provided by ETAK on licence to the OS. This routine will allow the algorithm to make decisions on road selection using road direction and access information. For example, on a dual carriageway, each carriageway is coded as a separate one-way road. A roundabout is also coded as a one-way road, as of course is a conventional one-way road. A simple comparison between a vehicle's direction of travel (its bearing) and the bearing of permitted travel along a road segment is used for this test.

Unfortunately, due to the high commercial value of the ETAK data, and the limited resources available at the time, only OSCAR Traffic Manager without DRI was available for the project, kindly supplied by the OS. The data structure used with OSCAR for DRI records was obtained from the OS web pages [EDINA, 2005] and replicated. The required DRI was collected manually for all roads in the region of the two routes and used to populate the replicated OSCAR Traffic Manager data records.

Before the development of these enhancements, the RRF based its entire map matching on road geometry. The RRF now has intelligence to select roads by network distance traveled and drive restriction information. This enhanced RRF, incorporating two new routines, is called Intelligent RRF (IRRF). The two new routines were also tested independently of each other; these versions of the algorithm are called RRF-network and RRF-DRI.

To evaluate the effectiveness of IRRF GPS C/A code observation data was collected in a vehicle driven on routes through urban Newcastle upon Tyne. Figure 6.7 (Plate 5) shows the two routes that were used. Route one is a fairly well built-up area and route two has good sky visibility. Over the same period, dual frequency phase data was collected in the vehicle and also by a static

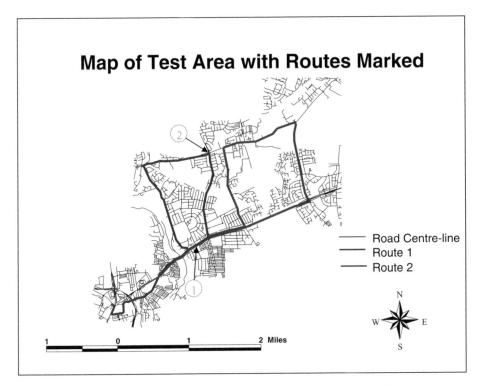

Figure 6.7 *Map of the test area of Newcastle showing the two routes that were used Reproduced with permission from the Ordnance Survey*

receiver recording base station data on the roof of the Department of Geomatics, University of Newcastle. This dual frequency data was used to compute a high precision (cm accuracy) GPS solution, which was assumed to be the 'true' position of the vehicle at each epoch (one second). All available satellites visible to both receivers were used in the position solution computation, this number varied throughout the route from none to eight. Six different types of point position solutions were computed for each route, see Table 6.4. RAW is the uncorrected stand-alone positions, which is the same as that output directly by the receiver. Height aiding was applied in solutions 2 to 5 using OS Digital Terrain Model (DTM) data, obtained from the EDINA Digimap service [OS, 2005].

Data Processing. All GPS data was transformed into RINEX format for processing all solutions, Gurtner [1993] provides an authoritative description of the RINEX format. The processing of the data for solutions numbers 1 and 6 was carried out using the GPS post-processing software GeoGenius. All other solutions were computed by processing the data with the various versions of the developed Road Reduction Filter algorithm. To correctly display the vehicle

Table 6.4 *Six point position solutions*

Solution type	Data
1 RAW	vehicle C/A code data
2 RRF	vehicle C/A code data, RRF, digital road and height data
3 RRF-Network	vehicle C/A code data, RRF with network tracing, digital road and height data
4 RRF-DRI	vehicle C/A code data, RRF with DRI data, digital road and height data
5 IRRF	vehicle C/A code data, IRRF, digital road and height data
6 RTK	dual frequency phase data from both the vehicle and the base station

positions on OS large-scale mapping, transformation software was developed as part of the work. The resultant latitude, longitude and height coordinates from the six solutions were transformed to OSGB36 (Ordnance Survey of Great Britain, 1936) National Grid, with a nominal transformation accuracy of 20 cm [OS, 1999]. At each epoch where all six position solutions were available, the difference in position between 6, the 'true' position, and each of the other five solutions were calculated (position error).

Results. On visual inspection, it could be seen that the original RRF selected the correct road segment when the vehicle was traveling along single carriageway roads. The largest proportion of road selection errors occurred where there were dual carriageways, slip roads and roundabouts. The RTK positions (assumed to be true) were good along all roads, as expected, but there were far fewer point positions. For both routes there was a total of 2039 points from the RRF calculation, but there was only 1060 RTK positions. This can be explained because to compute an RTK position at least five satellites are required, but because of height aiding from DTM data, all RRF solutions (2, 3, 4 and 5) can be computed with a minimum of three satellites visible.

RRF-DRI. The addition of the DRI routine has improved the road selection process. Route 2 in particular was improved unquestionably both through visual inspection and statistically. It can be seen on Figure 6.8 that there were improvements often in the order of about 1.5 metres.

RRF-Network. RRF plus a network shortest path algorithm also improved reliability for positioning a moving vehicle. Table 6.5 displays the differences in Easting (DE), Northing (DN) and position (DP) between the RTK and RRF solutions for route 1, this is the error from the 'true position'. Similarly Table 6.6 displays these differences for the RTK and the RRF-Network solutions. It can be seen from the tables that there are small changes in the mean and standard deviation, with mean position improving by approximately 0.3 m. However, removing gross errors (Figure 6.9) will further improve these figures. Further tuning of the RRF for network distance testing will possibly achieve this.

There is still some potential for further improvements using this approach. A fully integrated IRRF solution requires that the network distance comparison

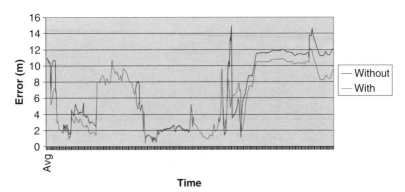

Figure 6.8 *Error in GPS positioning on route 2, with and without DRI processing*

Table 6.5 *Differences between RTK and RRF solution, route 1*

	DE	DN	DP
Mean	1.92	3.58	7.38
Standard deviation	3.95	6.11	3.87

Table 6.6 *Differences between RTK and RRF-Network solution, route 1*

	DE	DN	DP
Mean	1.54	3.96	7.19
Standard deviation	3.90	5.84	3.95

parameters are automatically adjusted by current velocity and acceleration of the vehicle. Furthermore, a full implementation of all Drive Restriction Information available, not only direction of permitted travel, would provide additional intelligence to road selection.

7. Conclusions

From Table 6.1 we can see that the position error can be resolved to within the expected random error of perfect DGPS for all except the slightest of change in

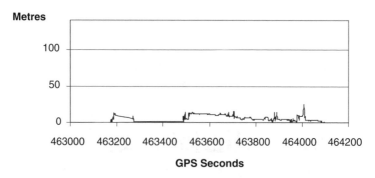

Figure 6.9   *Large position errors in route 1*

road geometry. Problems begin to arise with roads, which curve by only 20 degrees within the period that the error is assumed to be constant (~30 sec for road navigation), although even 10 degrees would be sufficient provided the bend is effectively instantaneous. We therefore conclude that only if roads are straighter than 10–20 degrees during a 30-second driving period (i.e., 0.4–1 km in typical driving conditions) will map-matched GPS be significantly worse than DGPS. However, the full precision of DGPS is certainly not required for finding the correct road centre-line, so these numbers are in any case extremely conservative for that purpose. In summary, we expect on firm theoretical grounds that combined RRF and map-matched GPS to be as good as DGPS for correct road centre-line identification in almost any possible circumstance. This has the distinct advantage of being a completely self-contained system, requiring no radio communication for differential corrections and continuous data provision. Furthermore, because the computation of the estimated GPS receiver position is part of the RRF and a digital terrain model-derived height aiding is used in the solution, only three satellites are necessary for a solution.

The main objectives of developing the two additional routines, one for network distance calculation and one to process Drive Restriction Information were to do the following:

- prove the feasibility of adding intelligence to the road selection process. This was achieved.
- improve the accuracy of point positions provided by the RRF map-matching algorithm. This has also been achieved, but not significantly.

7

The Use of Digital Terrain Models to Aid GPS Vehicle Navigation

This chapter describes further developments of the RRF map-matching algorithm. The potential improvements in reliability and accuracy of GPS vehicle positioning, when using different resolution digital terrain models (DTM) for height determination are discussed. Height-aided solutions are computed using a variety of commercially available DTMs, and a very high resolution DTM, with height accuracy of 40 mm, created as part of the work. The use of a higher order interpolant (e.g. a bicubic or biquintic polynomial) are investigated for its potential to improve performance, compared to a simple bilinear interpolant. This experiment demonstrates that when the number of satellites visible to the receiver is reduced, or the satellite geometry is poor, map matching and height aiding considerably improve the horizontal and elevation accuracy.

1. Introduction

Various map-matching techniques have been developed, such as the Road Reduction Filter (RRF) (see Chapter 5 for details). RRF tracks a vehicle on all possible roads in a computed error region, then uses a method of rapidly detecting, and filtering out, inappropriate road centre-lines from the set of all those possible. RRF processes raw GPS code data using its own least squares position estimation algorithm, which includes height aiding from digital terrain models (DTM). Height aiding can augment a GPS code solution, it can reduce the number of pseudorange observations required to obtain a solution, thus observations from only three satellites can provide a 3D position. RRF calculates its own GPS position error correction, using road geometry, in a formal least squares process (see Chapter 6 for details).

Intelligent Positioning: GIS-GPS Unification G. Taylor and G. Blewitt
© 2006 John Wiley & Sons, Ltd

In the experiments described in this chapter, heights are obtained from Ordnance Survey (OS) 50 m and 10 m gridded DTMs (1:50,000 scale LandForm Panorama and 1:10,000 scale LandForm Profile respectively). GPS points are unlikely to correspond to the grid height points in any DTM, therefore elevations for height aiding for GPS point position computation must be interpolated from the DTM. There are a variety of interpolation algorithms (e.g. linear, bilinear, bicubic and biquintic), which can be used to obtain heights from a DTM [Dorey, 2002; Kidner, 2003]. In many instances, linear techniques can produce a diverse range of solutions for the same interpolated point. A higher order interpolant that takes account of the neighbouring vertices, either directly or indirectly as slope estimates, will always produce a better estimate than the worst of these linear algorithms [Kidner *et al.*, 1999]. The more complex an interpolation algorithm is, the more computationally expensive it becomes, which may be a prohibitive overhead when computing GPS positions at high rates, such as ten point positions per second. The work described here investigates how the choice of DTM and interpolation method affects the RRF map-matching algorithm, in terms of horizontal position and height accuracy.

2. Digital Terrain Models

Digital terrain models (DTMs) are an important data source for geographical information processing. They are used to represent, analyse and visualize phenomena related to topography or other surfaces. A DTM may be understood as a digital representation of a portion of the earth's surface [Weibel and Heller, 1991]. Applications of digital terrain modelling abound in civil engineering, landscape planning, military planning, aircraft simulation, radio communications planning, visibility analysis, hydrological modelling, and more traditional cartographic applications, such as the production of contour, hill-shaded, slope and aspect maps [Kidner *et al.*, 1999]. There are various definitions of DTMs, and some authors argue that the term 'Digital Elevation Model' (DEM) should be used instead of 'Digital Terrain Model', where merely relief is represented. The term terrain often implies attributes of a landscape other than the altitude of the land surface [Burrough, 1986]. In contrast, Ordnance Survey names their products 'DTM' (which includes contours, break lines, etc.). They provide a definition which is specific only to their own products.

The accuracy of a DTM is an important issue, which is an essential factor for many GIS applications, for example, intervisibility and hydrological modelling. Often, existing topographic maps have formed the main data source for generating DTMs. Any DTM/DEM derived from digitizing a topographic map is an approximation of the real world [Carter, 1988]. Dorey [2002] suggests that small source errors can propagate through to large errors in such terrain models and inevitably in the final application itself. Since OS DTMs are used as the only experimental dataset in this work, it is necessary to evaluate the accuracy and suitability of such data. The OS have produced two DTMs at different grid resolutions (50 m and 10 m). The following three statements are made by the Ordnance Survey (2001) in terms of their DTMs' accuracies:

1 The height accuracy of any point in the DTM is equal to or better than half the contour interval, i.e. ±2.5 m for areas with 5 m vertical intervals and ±5 m for areas with 10 m vertical intervals.
2 The process of creating DTMs utilizes all the height information contained in the contour file to generate the height of each of the points in the DTM. The results achieved depend on the density of the height data contained in the contour file.
3 In some flat areas where there is little height information, contours and spot heights may be a great distance apart; this can cause irregularities in the DTM, which appear as slight terracing of the terrain.

These three statements have been thoroughly analysed by Dorey [2002]. The conclusion drawn in Dorey's work is that error in OS DTMs can often exceed specified tolerances of half the contour interval. Furthermore, the OS method of deriving DTMs using digitized contours, although relatively economic in terms of other available data sources, is intrinsically flawed and the quality of derived DTMs is not of an exceptionally high standard.

In the context of this experiment, elevation accuracy is one critical factor that will affect the accuracy of GPS positions obtained in a moving vehicle. There are other factors, e.g. satellite constellation geometry, number of satellites, etc. A variation in the error of a vehicle position, related to interpolated heights from OS DTMs, will be identified in the developed map-matching software.

3.   Spatial Interpolation of Elevation Data

In digital terrain modelling, interpolation serves the purpose of estimating elevations in regions where no data exist [Weibel *et al.*, 1991]. Throughout the literature, there are a variety of interpolation procedures, [Kidner *et al.*, 1999] which are widely used in different applications of GIS, for example, interpolation from contour lines or survey coordinates to generate a grid DTM. A given GPS coordinate point is unlikely to correspond to the measured height point in a DTM. In this experiment, elevations of GPS points collected by an autonomous receiver must be interpolated from a regular grid DTM. However, there are a large variety of point interpolation algorithms that can be used to obtain a height from a DTM for the purpose of height aiding. A study made by Dorey [2002] demonstrates that when interpolating DEMs, irrespective of terrain complexity, the higher order algorithms consistently outperform the simpler linear variant. For this study, two representative high order interpolation algorithms, bicubic and biquintic are implemented, as well as the more popular bilinear algorithm, often incorporated in desktop GIS packages.

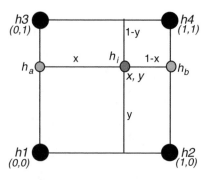

Figure 7.1   Bilinear interpolation

3.1.   Patchwise Polynomial Interpolation

The most commonly used interpolation method for a regular grid is patchwise polynomial interpolation. The general form of this equation for surface representation is:

$$h_i = a_{00} + a_{10}x + a_{01}y + a_{20}x^2 + a_{11}xy + a_{02}y^2 + a_{30}x^3 + a_{21}x^2y + a_{12}xy^2 + a_{03}y^3$$
$$+ a_{31}x^3y + a_{22}x^2y^2 + a_{13}xy^3 + a_{32}x^3y^2 + a_{23}x^2y^3 + a_{33}x^3y^3 + \ldots + a_{mn}x^my^n$$

where h_i is the height of an individual point i, x and y are the rectangular coordinates of i, and $a_{00}, a_{10}, a_{01}, \ldots, a_{mn}$ are the coefficients of the polynomial.

The bilinear polynomial can be represented as follows, using a local unit square coordinate system whose origin is $(0, 0)$:

$$h_i = a_{00} + a_{10}x + a_{01}y + a_{11}xy$$

By substituting the coordinates of the four vertices indicated in Figure 7.1 into this equation, then the coefficients can be solved such that:

$$h_i = h_1 + (h_2 - h_1)x + (h_3 - h_1)y + (h_1 - h_2 - h_3 + h_4)xy$$

where h_i is the height of the point to be interpolated.

3.2.   Bicubic Interpolation

Bicubic interpolation makes use of the 16-term function:

$$h_i = a_{00} + a_{10}x + a_{01}y + a_{20}x^2 + a_{11}xy + a_{02}y^2 + a_{30}x^3 + a_{21}x^2y + a_{12}xy^2$$
$$+ a_{03}y^3 + a_{31}x^3y + a_{22}x^2y^2 + a_{13}xy^3 + a_{32}x^3y^2 + a_{23}x^2y^3 + a_{33}x^3y^3$$

The values of the first derivative in each direction (i.e. f_x and f_y), the cross-derivative (f_{xy}), and the elevations at each of the four grid vertices amount to 16 known values. Since there are 16 unknown coefficients (i.e. a_{mn}), these can be solved simultaneously. However, there are a variety of methods for estimating the various derivatives at the grid vertices, with some being better than others [Kidner, 2003].

3.3. Biquintic Interpolation

Biquintic interpolation makes use of the 36-term patchwise polynomial:

$$h_i = \sum_{i=1}^{6} \sum_{j=1}^{6} a_{ij} x^{i-1} y^{j-1}$$

In order to solve the 36 unknown coefficients, the derivatives $f_x, f_y, f_{xx}, f_{xy}, f_{yy}, f_{xxy},$ f_{xyy}, and f_{xxyy} and the elevations at each of the four grid vertices are required to generate a 36×36 system of equations. For the bicubic and biquintic interpolation algorithms, finite difference approximations are used for the purpose of calculating the partial derivatives [Kidner, 2003].

4. Map Matching and the Road Reduction Filter

Scott's [1994] mathematical framework for map matching of positions using GPS (see Chapter 5, Section 2), acknowledges that a key component of the map-aided estimator is correct road identification. The fact that users are generally constrained to a finite network of paths or roads provides computer algorithms with digital information that can be used to correlate the computed location with the digitized network. This is known as map matching [White, 1991]. Many methods have been devised for map matching [e.g., Scott, 1994; Mallet *et al.*, 1995; and Taylor *et al.*, 2001a, 2001b]. We now present the development and testing of an algorithm that utilizes GPS for the initial position, and geometric information, computed from the digital network itself, as the only other source of data for map-matching.

4.1. Road Reduction Filter (RRF)

A novel method of map matching using GPS has been developed which uses digital mapping and height data to augment point position computation. Chapter 5 describes the RRF in full detail. This Road Reduction Filter (RRF) method reduces the error in position, which is a sum from several sources, including signal delay due to the ionosphere and troposphere. The general approach adopted in the development of RRF for GPS receiver tracking is initially to identify all possible candidates for the correct road or path on which the receiver is travelling on, and then to systematically remove the wrong ones.

All candidate road segment arcs are compared with the trajectory described by the uncorrected RAW GPS point positions. This comparison is based on the distance travelled and bearing of the RAW point positions compared with the corresponding positions on road centre-lines. Furthermore, height aiding (HA) is used in the position computation. HA uses height data, obtained from the DTMs. An enhancement to RRF, using a formal method of computing a map matched correction, is integrated into the Road Reduction Filter. The incorporation of this error vector into RRF is detailed in Chapter 6. This map-matched correction, or 'error vector', is used to adjust the position of the GPS receiver on the road segment, but only when the error vector is known to be of high quality (residual values are low). This error vector is calculated using a least squares estimation, based on the last n receiver positions, where n may range from as few as 5 epochs to 30 or more, depending on the road geometry. The advantage of formal methods is that quality measures can be derived and used to place confidence bounds for rigorous decision-making (for example, to reject road centre-lines that fail a particular hypothesis test). Formal methods also provide an insight into the relative importance of factors, which can improve the procedure (e.g. data rates and road geometry).

5. Data Collection and Processing

To evaluate the effectiveness of the RRF map matching algorithm, GPS Coarse Acquisition (C/A) code observation data was collected in a vehicle driven on roads in the hills of rural Glamorgan, Wales, see Figure 7.2. Simultaneously, GPS phase data observations were collected in the vehicle and also by a static receiver recording base station data, on the roof of a nearby building at the University of Glamorgan (i.e. within 5 km). This phase data was used to compute a high

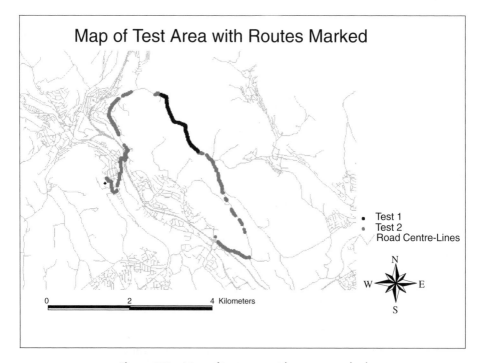

Figure 7.2 *Map of test area with routes marked*

Table 7.1   *Experimental data sets*

Data set	GPS receiver	Method of surveying
Test1	SR510 L1 single frequency, code and phase measurements	Kinematic, (post-processing)
Test2	SR520 L1 and L2 dual frequency, code and phase measurements	Kinematic on the fly ambiguity resolution, (post-processing)
Test3	SR510 L1 single frequency, code measurements only	DGPS, (post-processing)

precision (cm accuracy) GPS solution, which was assumed to be the 'true' position of the vehicle at each epoch (one second).

All available satellites visible to both receivers were used in the position solution computation. This number varied throughout the route from none to ten. Four point position solutions were computed:

RAW solution	using C/A code data
RAW + HA solution	using C/A code data and interpolated height
RRF map-matched GPS solution	using C/A code data, interpolated height, RRF with digital road map data.
RTK solution	using single and dual frequency phase data from both the vehicle and the base station to compute a high precision (cm accuracy) GPS solution. The 'true' position of the vehicle at each epoch was assumed to be that given by this solution.

Three data sets were collected for this experiment, (see Table 7.1), but only the results of test 1 are discussed in detail in this chapter. Data collection was accomplished using Leica 500 series geodetic GPS receivers. Phase data processing was achieved with Leica Ski-Pro software. All coordinates were transformed from WGS-84 (World Geodetic System, 1984) to OSGB36 (Ordnance Survey of Great Britain, 1936) National Grid.

5.1.   Accuracy of Solution

The accuracy of the RTK solution [Leica, 1999] was 10–20 mm in plan and 20–40 mm in height. A summary of the total number of processed epochs (1 second) is given in Table 7.2. If the RTK positions are assumed to be the true position, the position errors are computed according to the method detailed in Chapter 5 (and summarized here):

$$DP = \sqrt{DE^2 + DN^2}$$

where DP is the position error, DE is the Easting error, and DN is the Northing error, while:

Table 7.2   *Processed epochs*

	RTK	RAW	RAW+HA	RRF
processed epochs of Test1	274	1267	1267	1267

Table 7.3   *Height error in test1 (RTK-RAW+HA)*

	RTK-RAW+HA						RTK-RAW
	50 m OS DTM			10 m OS DTM			N/A
	Bilinear	Bicubic	Biquintic	Bilinear	Bicubic	Biquintic	N/A
Mean	−18.291 m	−18.271 m	−18.272 m	−17.656 m	−17.653 m	−17.653 m	−22.413 m
SD	2.131 m	2.092 m	2.092 m	1.946 m	1.947 m	1.947 m	2.314 m
Minimum	−20.631 m	−20.660 m	−20.573 m	−18.888 m	−18.902 m	−18.903 m	−24.034 m
Maximum	−9.376 m	−9.438 m	−9.442 m	−8.587 m	−8.566 m	−8.564 m	−11.889 m

$$DE = E_{RTK} - E_i$$
$$DN = N_{RTK} - N_i$$

where E_{RTK} and N_{RTK} are the Easting and Northing of the RTK position, and E_i and N_i are the corresponding positions for each of the other methods (i.e. RAW, RAW+HA, and RRF).

6.   Results

As illustrated in Table 7.3, a comparison of height errors in terms of the mean, standard deviation, minimum and maximum error when computing RAW+HA and those when computing a RAW (single GPS receiver) is made.

6.1.   Height Errors – Test 1

These results demonstrate that there is an improvement of at least 4 m in height accuracy when using height aiding. However, these results also demonstrate that there is no significant difference (less than 0.02 m) between the mean height errors of the three interpolation algorithms when using the same resolution DTM. However, using interpolated height values from the higher resolution (10 m) DTM gives more accurate results (about 0.6 m) than the 50 m DTM in terms of the mean height error. This is to be expected due to the larger source scale and greater accuracy of the 10 m DTM.

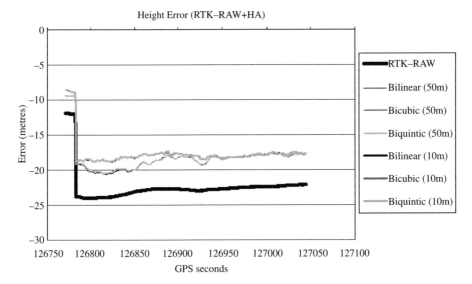

Figure 7.3 *Height error in test1 (RTK-RAW+HA)*

Table 7.4 *Position error in test1 (RTK-RAW+HA)*

	RTK-RAW+HA						RTK-RAW
	50 m OS DTM			10 m OS DTM			N/A
	Bilinear	Bicubic	Biquintic	Bilinear	Bicubic	Biquintic	N/A
Mean	5.051 m	5.046 m	5.046 m	4.916 m	4.916 m	4.916 m	5.890 m
SD	0.595 m	0.595 m	0.595 m	0.590 m	0.590 m	0.590 m	0.477 m
Minimum	4.680 m	4.689 m	4.689 m	4.617 m	4.607 m	4.606 m	5.616 m
Maximum	7.735 m	7.739 m	7.739 m	7.673 m	7.675 m	7.675 m	8.047 m

Figure 7.3 also shows that the OS 10 m DTM provides more accurate interpolated heights than the OS 50 m DTM, before epoch 126950. After this epoch,
the two DTMs provide very similar results, as the terrain was much more varied
at the start of the route. Again, no significant difference between the height
errors of the three interpolation algorithms is shown in Figure 7.3. With regard
to height errors of a single GPS receiver (RAW), Figure 7.3 demonstrates that
the height errors provided by RAW are consistently larger (about 4–5 m) than
those provided by RAW+HA. Therefore, height aiding does give us more accurate heights than a single GPS receiver, even though the interpolated height itself
is calculated at a coordinate point on average 6 m from the receiver's true location, see Table 7.4. This has been clearly shown in Figure 7.3.

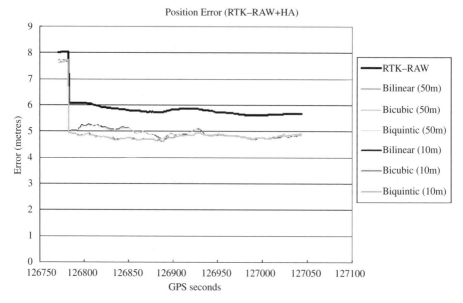

Figure 7.4   *Position error in test1 (RTK-RAW+HA)*

6.2.   Position Errors – Test 1

However, height errors provided by RAW+HA are still large (15 m–20 m), because of the inaccurate raw GPS positions. Table 7.4 shows the calculated position error. In the relatively rugged terrain where the data were collected, i.e. the hills of South Wales, any error in the computed receiver horizontal position may result in a large interpolated height error. This is because hill slope in the vicinity of these road surfaces may be 45 degrees or higher. This slope can give a very rapid change of height any distance away from the road. Even if the height aiding is used alone (i.e. RAW+HA), all the interpolated heights are still obtained from the road. RRF will always try to snap the raw position onto the road centre-lines to obtain a more accurate interpolated height.

As shown in Table 7.4, there is no significant difference (less than 0.01 m) between the mean position errors of the three interpolation algorithms when using the same resolution DTM. The larger resolution 10 m DTM provides slightly more accurate positions (less than 0.14 m) than the 50 m DTM in terms of mean positional error. In addition, it is clear (see Figure 7.4) that height aiding does improve the positioning accuracy of a single receiver (i.e. about 1 m improvement in horizontal accuracy).

In order to test the overall performance of RRF, both OS road centre-line data and OS DTM data are used to compute an RRF position (Table 7.5). Here, the height error is reported as a general Root Mean Square Error (RMSE), which is given by:

$$RMSE = \sqrt{\frac{\sum_1^n (DH)^2}{N}}$$

As shown in Table 7.5, the high resolution 10 m DTM significantly outperforms the 50 m DTM in terms of RMSE. Further analysis of these height errors (Figure 7.5) shows a similar, but exaggerated trend to the RAW+HA results. On the hilly roads, the errors were more pronounced between the different resolution DTMs, but after epoch 126950 (in the flatter terrain), the discrepancies were less noticeable. However, it is clear that height errors provided by RAW and RAW+HA are consistently larger than those provided by RRF.

This can be explained because positions corrected by RRF are snapped onto road centre-lines, which will reduce the height error as the true receiver (RTK)

Table 7.5   *Height error in test1 (RRF)*

	RTK-RRF						RTK-RAW
	50 m OS DTM			10 m OS DTM			N/A
	Bilinear	Bicubic	Biquintic	Bilinear	Bicubic	Biquintic	N/A
RMSE	4.258 m	4.026 m	4.032 m	0.799 m	0.828 m	0.829 m	22.532 m
Minimum	−9.151 m	−8.801 m	−8.836 m	−1.533 m	−1.579 m	−1.581 m	−24.034 m
Maximum	1.432 m	1.703 m	1.717 m	2.466 m	2.686 m	2.702 m	−11.889 m

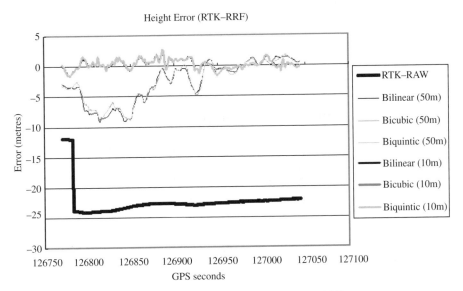

Figure 7.5   *Height error in test1 (RTK-RRF)*

positions are always on the road. Furthermore, there is an indication in Figure 7.5 that the terrain is best represented in the larger resolution 10 m DTM, particularly in the area of steeper slopes (i.e. in the first half of the data collection route). In addition, in the case of the 50 m DTM, the biquintic and bicubic are slightly better, by about 0.23 m, than the bilinear in terms of RMSE. However, in the case of the 10 m DTM, the bilinear is conversely better (about 0.03 m) than the biquintic and bicubic. So, there is no significant difference (less than 0.24 m) in height error between the three interpolation algorithms when interpolating from the same resolution DTM. As expected, the RMSE of RAW (22.5 m) is the worst compared with RRF.

Table 7.6 shows the position errors provided by RRF and those provided by RAW. It is very clear from Table 7.6 and Figure 7.6 that RRF provides a more

Table 7.6   *Position error in test1 (RRF)*

	RTK-RRF						RTK-RAW
	50 m OS DTM			10 m OS DTM			N/A
	Bilinear	Bicubic	Biquintic	Bilinear	Bicubic	Biquintic	N/A
Mean	3.963 m	3.245 m	3.245 m	3.247 m	3.248 m	3.248 m	5.890 m
SD	1.871 m	1.579 m	1.579 m	1.548 m	1.548 m	1.548 m	0.477 m
Minimum	0.345 m	0.329 m	0.329 m	0.344 m	0.343 m	0.343 m	5.616 m
Maximum	7.657 m	6.335 m	6.335 m	6.338 m	6.338 m	6.338 m	8.047 m

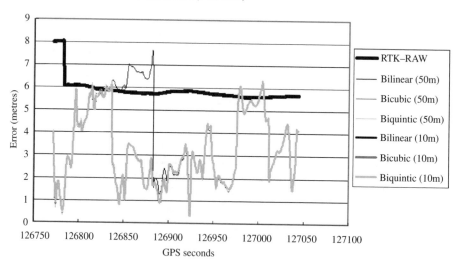

Position Error (RTK–RRF)

Figure 7.6   *Position error in test1 (RRF)*

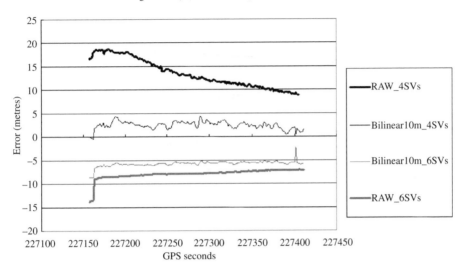

Height Error (RTK–RAW+HA)

Figure 7.7   *Height error in test2 (RAW+HA)*

reduced position error than RAW, the improvement from the mean position error is about 2.6 m except for the bilinear interpolation from the 50 m DTM. In addition, the RRF positions are also more accurate than RAW+HA positions in terms of the mean position error.

As illustrated in Figure 7.6, the bilinear interpolation from the 50 m DTM has much larger position errors (6–7.5 m) between epochs 126837 and 126882 than the other interpolation algorithms. This may be due to the height RMSE of the bilinear interpolation from 50 m DTM being high (see Table 7.5). The use of interpolated heights, with large errors, for GPS height aiding in RRF could cause large position errors. However, all the curves in Figure 7.6, except the bilinear (50 m) and RAW, are almost identical. This means that the two different resolution OS DTMs and the three interpolation algorithms have nearly the same effect on the positioning accuracy of RRF.

It is difficult to draw any substantial conclusions from the analysis of results obtained from only one data set. A second data set (test2) was collected using a dual frequency GPS receiver, which was used to confirm the results obtained in test1 (see Figure 7.2). For simplicity and clarity, all other curves describing the effects caused by the two high order interpolation algorithms, and different DTMs, in terms of the position error and height error, were removed from Figure 7.7 and Figure 7.8. The results obtained in test2 are nearly identical to those obtained from test1. As illustrated in Figure 7.7 and Figure 7.8, the height accuracy of raw data in test2, which is represented by the lowest curve ranging between –15 m and –5 m in Figure 7.7, has an average height error of about –10 m. This height error is much smaller than –22.4 m from test1 (see Table 7.3

Position Error (RTK–RAW+HA)

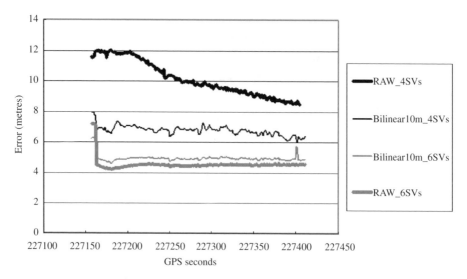

Figure 7.8   Position error in test2 (RAW+HA)

and Figure 7.5). Due to the high accuracy of the raw height in test2, height aiding alone may no longer improve the horizontal positioning accuracy of a single GPS receiver – this is clearly shown in Figure 7.8. The position error of raw data in Figure 7.8 has a constant error of 4.5 m. Therefore the position error provided by RAW+HA (about 6 m) is obviously less accurate than that provided by raw data in test2.

In order to further examine the results, another experiment was conducted. In this experiment, full control of which satellites are used to calculate a point position in the least squares estimation is provided. As a general rule (see Chapter 2), the larger the constellation of satellites, the better the available geometry, i.e. the lower the position dilution of precision (PDOP) (Sickle, 1996). High PDOP indicates poor satellite geometry and possibly less accurate point positions. In the context of this experiment, the azimuth and the elevation of each satellite were measured to give a clear picture of the distribution of satellites available to calculate a vehicle position.

7.   Results from Test2 Data with a Subset of Satellites

A subset of satellites was selected and then used in different combinations to demonstrate the effectiveness of RRF under simulated poor satellite geometry. Simply switching off certain satellites will significantly increase the value of PDOP and therefore result in large height and position errors in the raw GPS data of test2, as shown in Figure 7.7 and Figure 7.8. For the full set of six

Table 7.7   *Position error in test2*

Solution	Mean position error
RAW 4 satellites	10.0 m
RAW+HA 4 satellites	6.8 m
RRF 4 satellites	4.1 m

satellites, the RAW position PDOP values were consistently 3, and with four reduced satellites, the values ranged from 11.1 to 16.7. There is a total of six satellites available in the raw data of test2. Removing satellites PRN8 and PRN27, leaving four satellites, will dramatically increase the position error and height error of raw data to 4 m to 8 m and 15 m to 25 m respectively. With these intentionally corrupted raw positions (shown in Figure 7.8 and Table 7.7), the position errors obtained by RAW+HA with four satellites are much smaller than those provided by RAW with a subset of four satellites.

7.1.   Position Error – Test2

In general, using OS DTMs for height aiding and RRF (snapping to road centrelines) can improve the horizontal and height accuracy of a low cost stand-alone GPS receiver, when satellite geometry is poor and/or when the number of satellites visible to the receiver is less than optimal.

8.   Conclusion

It has been demonstrated that height aiding always improves the height provided by a single GPS receiver (RAW). RAW+HA in test1 provides a height improvement of about 4.1 m to 4.8 m. This is despite the position (coordinate) for the interpolated height being in error. However, map matching (RRF) and height aiding provide more significant height improvements (18.5 m to 21.7 m) than those provided by height aiding alone. With regard to the choice of interpolation algorithms, no algorithm is found to be significantly superior to others in terms of the height error when using map matching. However, irrespective of interpolation algorithms, interpolation from the larger resolution 10 m DTM consistently provides more accurate heights than the 50 m DTM. As would be expected, the nature of the terrain itself is highly significant in differentiating interpolation accuracies between different DTMs.

With regard to the horizontal positioning accuracy (i.e. Easting and Northing), height aiding can provide a horizontal position improvement of about 1 m compared with the position provided by a single GPS receiver in test1. The height errors of a single GPS receiver in test1 are significant, i.e. the mean height error is up to 22.4 m. The results suggest that if DTMs for height aiding alone are used to improve the positioning accuracy of a single GPS receiver, horizontal accuracy improvement depends on GPS receiver height accuracy, which itself depends on plan position accuracy. However, if both road centre-line data and DTM data are used to compute the RRF position (height aiding and map matching), RRF consistently improves the horizontal positioning accuracy of a single GPS receiver, typically by about 2 m. In addition, as far as the overall performance of map matching is concerned, the two DTM resolutions and three interpolation algorithms have nearly the same effect on the positioning accuracy of map-matching algorithm. That is, the accuracy and reliability of a map-matching algorithm, such as RRF, do not vary significantly, when using any combination of the two different resolution DTMs and the three different interpolation algorithms.

Overall, the use of the higher resolution DTM significantly improved the height accuracy of both the height aiding and map matching with height aiding approaches, but had very little benefit in improving the horizontal position. The use of higher order interpolation algorithms produced a noticeable, but small improvement in results for the lower resolution DTM, but no improvement for the higher resolution DTM. This suggests that at the smaller scale, the use of bicubic or biquintic polynomials can compensate for the poorer quality terrain data by identifying the relevant trends (i.e. slopes). However, with the larger-scale DTM, the denser sampling of data negates the need for a better interpolation algorithm. As the trend today is for higher resolution, gridded DTMs, then bilinear interpolation is satisfactory.

Furthermore, when the number of satellites visible to the receiver is reduced, or the satellite geometry is poor (high value of PDOP), map matching and height aiding, using OS DTMs, will considerably improve the accuracy of both plan position and height. The further experiment with test2 data proved this.

8

GPS Accuracy Estimation Using Map-Matching Techniques: Application to Vehicle Positioning and Odometer Calibration

This chapter describes the further development of the test-bed application in Chapters 5 and 6, called Map-Matched GPS (MMGPS) that processes raw GPS output data from RINEX files or GPS-derived coordinates. This developed method uses GPS point positioning, which is map matched to locate the vehicle on a road centreline when GPS is known to be sufficiently accurate. MMGPS software has now been adapted to incorporate positioning based on odometer-derived distances (OMMGPS), when GPS positions are not available. Relative GPS positions are used to calibrate the odometer. If a GPS position is detected to be inaccurate, it is not used for positioning, or for calibrating the odometer correction factor. In OMMGPS, GPS pseudorange observations are combined with DTM height information and odometer positions to provide a vehicle position at 1-second epochs. The described experiment used GPS and odometer observations taken on a London bus on a pre-defined route in central London. Therefore, map-matching techniques are used to test GPS positioning accuracy, and to identify grossly inaccurate GPS positions. In total, over 15,000 vehicle positions were computed and tested using OMMGPS.

In general, the position quality provided by GPS alone was extremely poor, due to multipath effects caused by the urban canyons of central London, so that odometer positioning was used much more often to position the vehicle than GPS. Typically, the ratio is 7:3 odometer positions to GPS positions. In the case of one particular trip, OMMGPS provides a mean error of position of 8.8m compared with 53.7m for raw GPS alone.

Intelligent Positioning: GIS-GPS Unification G. Taylor and G. Blewitt
© 2006 John Wiley & Sons, Ltd

1. Introduction

Global navigation satellite systems (GNSS), such as GPS, have been increasingly used in real-time tracking of vehicles. Especially when GPS is integrated with increasingly powerful geographic information system (GIS) technologies, the accuracy and reliability of low-cost stand-alone GPS receivers can be significantly improved to meet the technical requirements of various transportation applications of GPS, such as vehicle navigation, fleet management, route tracking, vehicle arrival/schedule information systems (bus/train) and on-demand travel information. With the autonomous European Satellite Navigations System Galileo, expected in 2008, an opportunity of a joint system 'GPS + Galileo' with more than 50 satellites will provide many advantages for civil users, in terms of availability, reliability and accuracy. However, severe multipath effects will continue to be a problem in dense urban areas.

To date, there have been many attempts to improve the reliability of vehicle positioning through the fusion of observations obtained by the integration of various positioning and navigation instruments. The vast majority of such systems use a GNSS, for absolute positioning, and a variety of other sensors to provide relative positioning. The usual model is to use GPS to position a vehicle whenever possible together with some form of inertial navigation system (INS) or dead reckoning (DR) system, such as odometer, gyro and compass, to determine a vehicle's position relative to an initial position.

Kealy *et al.* [1999] and Ramjattan and Cross [1995] describe a typical solution, integrating GNSS and DR using a Kalman filtering technique. In this experiment a test route for the system was established in the centre of Perth, Western Australia. The results of this work found that 'DGPS/DR solution starts to degrade from 1 m to errors as much as 35 m by the end of a 10 minute period' [Kealy *et al.*, 1999]. Kalman filtering techniques do have an inherent problem, for vehicle navigation, on road networks, 'in terms of stability, computational load, immunity from noise effects and observability' [Chiang *et al.*, 2002]. The performance of the filter is heavily dependent on the models used. The model used is a compromise between a statistical predictive dynamic model and the measurement (observation) model. If too much weight is given to the dynamic model, an overly smooth track is the result, i.e. rapid changes of direction are not recognized quickly enough. If too much weight is given to the measurement model, errors would be construed as sharp changes in direction. Devising the correct model is very difficult, and without a very good model a Kalman filter will deliver the wrong result. Other accounts of using a Kalman filter for multi-sensor vehicle navigation are given by Stephen and Lachapelle [2000], who use GPS and low cost gyro, by Petrovello *et al.* [2003], who provide an informative discussion on levels of integration, and also by Mezentsev *et al.* [2002]. Hailes [1999] uses Kalman filtering with map matching.

Fei *et al.* [2000] describe fuzzy logic techniques as an alternative to Kalman filtering for GPS/INS integration. Furthermore, Mayhew and Kachroo [1998] compare solutions using various configurations of GPS, steering position, odometer, gyroscope, forward accelerometer and map-matching, with sensor fusion methods Kalman filtering, rule-based and fuzzy logic. Chiang *et al.* [2002] developed a GPS/INS multi-sensor navigation system that utilizes an artificial neural network (ANN) as another alternative to Kalman filtering. Wise-McLain and Murphy [1993] describe GPS and a DR system for tracking.

Over the past four years a group of researchers from the GIS Research Centre, School of Computing, University of Glamorgan, have designed, developed and implemented a software application package for researching algorithms and techniques to improve GPS based on map matching for navigation and tracking. This test-bed application, called Map-Matched GPS (MMGPS) processes raw GPS output data, from RINEX files, or GPS-derived coordinates. It provides linkage to a GIS for access and analysis of appropriate spatial and related attribute data (primarily road and height information). MMGPS identifies the correct road, on which a vehicle is travelling, and snaps the vehicle position onto that road. Furthermore, MMGPS corrects the derived position using its own computed correction parameters, e.g. Correction Dilution of Precision (CDOP) using the history of previous position estimates and road geometry [Blewitt and Taylor, 2002]. Various research experiments utilizing MMGPS have been conducted and results have been fully described in Taylor *et al.* [2001a].

The main objectives of the research described in this chapter are to determine both the accuracy and reliability of position of a public transport bus that can be provided using a combination of GPS, odometer and map matching techniques. To this end, a new algorithm has been developed that integrates odometer observations with the existing software, now called OMMGPS. In OMMGPS, height information obtained from digital terrain models (DTM) are used to determined 3D GPS point positions, when only three GPS satellites are visible to the receiver. More importantly, height aiding improves the accuracy of GPS point positions with poor satellite geometry (high PDOP), and when severe multipath occurs (multiple reflected GPS satellite signals).

The developed method uses absolute GPS positioning, map matched, to locate the vehicle on a road centre-line, when GPS is known to be sufficiently accurate. When this is not the case, odometer readings are used to locate the vehicle on a road centre-line. The odometer is calibrated using relative GPS positions, based on map-matching criteria, such as the residuals of CDOP, for GPS precision determination (see also Section 2). The accuracy of OMMGPS is a function of the frequency of accurate GPS points, reliable map matching and correct odometer calibration.

Standard Ordnance Survey (OS) digital plan and height map products were used for road map matching and height aiding. A number of trips along a bus route in central London – Baker Street, Oxford Street, etc. – were made to test the method. A typical result of map-matched GPS positioning is shown in Figure 8.1.

The innovative feature of this particular implementation is that map matching is actually used for GPS accuracy determination rather than to identify the correct road the bus is driving on. This is achievable, since pre-defined bus routes

London

- MMGPS
- Uncorrected GPS
  Osnames.shp
- Route2a_offset.shp
- Water_polyline.shp
- Road_polyline.shp
- Rail_polyline.shp
  Names_text.shp
- General_polyline.shp
- Fence_polyline.shp
- Building_polyline.shp
- Boundary_polyline.shp

0.09   0   0.09   0.18 Miles

Figure 8.1   *Map-matched GPS positioning with OMMGPS*

are involved, hence the correct road is always known. The trajectory of a sequence of GPS point positions is compared with an identified part of the bus route, using the map-matching techniques described in this chapter. If this comparison meets the map-matching criteria, described below, the points are used for odometer calibration and bus positioning. Otherwise, they are discarded.

2.   Methodology

The general approach was to use GPS to position the vehicle, and also to calibrate the odometer readings, but only if GPS was available and of sufficient accuracy. At all other times odometer readings were used to position the bus. Odometer positioning was achieved by tracing the distance measured by the odometer along the bus route road centre-line – actually, a 5 m offset center-line was used, left of direction of travel, see Figure 8.1. Map-matching techniques are used to improve the GPS positioning accuracy, and to identify grossly inaccurate GPS positions. The previous 10 GPS/odometer positions are used for map-matching calculations.

3.   Map Matching

The existing MMGPS software has been adapted to incorporate positioning based on odometer-derived distances, when GPS positions are not available. This new version of map-matching software, OMMGPS, works in the following way:

1 A GPS observation is read from the GPS RINEX file, and an odometer count is read from the odometer file.

2 A **Raw** vehicle position is computed using all satellites visible to the receiver, above a 15 degrees elevation mask, plus height aiding, where height is obtained from a DTM. This height is interpolated (bilinear) at the vehicle's previous reference (**Ref**) position, i.e. snapped on the road centre-line [Li *et al.*, 2003]. This DTM-derived height of the receiver is used to provide an extra equation in the least squares pseudorange computation of GPS coordinates, i.e. computation with a minimum of three satellites is possible. For each epoch (instant time of observation) GPS points within 100 m of the road centre-line are considered.

3 An odometer correction is calibrated using odometer count at the current epoch and relative GPS distance travelled. Only if the GPS position is usable, otherwise this step is skipped for this epoch.

4 Road geometry based on DGPS corrections (from step 9, previous epoch) is added to the **Raw** position to give a corrected (**Cor**) position.

5 This **Cor** position is now snapped to the nearest point on the nearest road-centre line to give the current **Ref** position, and then a DTM height is calculated. That is, a **Ref** position is available that can be used to generate road geometry based on DGPS corrections for use with the next epoch's computed **Raw** position. The resultant **Ref** positions are checked for correctness using tests against map-matching criteria [Taylor and Blewitt, 1999, 2000; Taylor *et al.*, 2001a]; see below.

6 The odometer position is calculated, using previous **Ref** position, as well as the odometer distance.

7 A position error vector is estimated in a formal least squares procedure, in which the Correction Dilution of Precision (CDOP) is computed. This estimate is a map matched correction that provides an autonomous alternative to DGPS, fully described in Chapter 6 and in Blewitt and Taylor [2002]. The residuals of this process are used to determine the goodness of fit of the position error vector.

8 Position error vector (step 7) is used to adjust the **Ref** position used for step 9. This provides a long-track correction, especially when CDOP is low (rapid change of road direction).

9 DGPS corrections for each satellite pseudorange are computed using the current **Ref** position. These are retained for future use.

The main map-matching criteria for snapping a **Raw** GPS position to a road centre-line **Ref** GPS position are each of the following, which have to be below a set maximum value:

- *Distance error* (absolute value of the difference between **Raw** distance and **Ref** distance, between the current and previous epochs);
- *Bearing error* (absolute value of the difference between **Raw** bearing and **Ref** bearing, between the current and previous epochs);

- *Residuals of CDOP*;
- *Maximum distance* of **Raw** GPS position from the road centre-line.

The number of satellites visible to the receiver has to be the same for current and previous epochs.

If a Ref GPS position passes the check, it is used for positioning the bus and calibrating the odometer correction factor. Otherwise, the position of the bus is derived from the calibrated odometer distance, and the odometer calibration correction factor is not updated. The values used for map matching are obviously open to adjustment and tuning, for different road geometry and environmental scenarios.

4.   Distance Correction Factor

The previous section assumes that distances obtained from odometer readings are multiplied by a *correction factor* C, so that the distance supplied to OMMGPS is actually Cd rather than just d. It is not reasonable to assume C is fixed, as different roads and, indeed, different road conditions on the same road will influence C. For this reason, when GPS and odometer signals are both available, C will be calibrated over a time window by comparing distances travelled, based on odometer readings with those estimated from GPS (i.e. relative distances between two GPS points). If GPS goes offline, the value of C obtained just before the GPS signal is lost (or regarded as unreliable) is used together with the odometer method to estimate location, i.e. the current odometer position is calculated based on the previous GPS or odometer position, and the current odometer reading is multiplied by the correction factor C.

5.   Estimating C

C can be regarded as a correction factor between odometer-based distance estimates as used above, and those obtained from the GPS-based method. At each second t, an odometer distance d_t and GPS-based coordinate estimates (X_t, Y_t, Z_t) are obtained. Here, it is assumed d_t is a cumulative variable, so that the distance travelled between $t-1$ and t is $d_t - d_{t-1}$. This is called Δd. Also, from the GPS measurements the cumulative distance travelled can be computed using OMMGPS. These distances are called D_t. Similar to the odometer distances, ΔD is defined as $D_t - D_{t-1}$. Thus, a model of the relationship between Δd and ΔD is:

$$\Delta D_i = C\Delta d_i + error$$

This model can be calibrated by estimating C using least squares techniques, i.e. C is chosen to minimize the expression:

$$\sum_i \left(\Delta D_i - C\Delta d_i\right)^2$$

It may be verified that, in this case, the estimate for C is:

$$C = \frac{\sum_i \Delta D_i \Delta d_i}{\sum_i \Delta d_i^2}$$

However, this assumes that C is a constant correction factor. In reality, C is likely to change, depending on traffic conditions, such as road shape, weather, and so on. A more realistic model allows C to vary with time, so that at each time t a distinct C_t is obtained. One approach in this situation is to estimate C according to the same model, i.e. using a 'moving window' least squares estimate. At each time t, only the values for ΔD_i and Δd_i are considered in a time window of k seconds, i.e. only data from times $t - k, t - k + 1, \ldots, t$. At time $t + 1$, data is dropped for time $t - k$ and added for time $t + 1$. Also, a weighting scheme is used in the least squares method, so that the squared errors for data close to t have a higher weighting. This gives an estimation method, which places more emphasis on minimizing errors close to time t. In this case, the least squares expression to be minimized is:

$$\sum_{i=0\ldots k} w_i \left(\Delta D_{t-i} - C_t \Delta d_{t-i} \right)^2$$

where w_i is the weight placed on the error at lag i seconds before time t. In this case, C_t is:

$$C_t = \frac{\sum_{i=0\ldots k} w_i \Delta D_{t-i} \Delta d_{t-i}}{\sum_{i=0\ldots k} w_i \Delta d_{t-i}^2}$$

5.1.   Weighting Scheme for w_i

Some thought should be given to the weighting scheme for the w_is. Obviously, a time-decay effect is expected, so that $w_0 > w_1 > \ldots > w_k$. One possibility is to choose an exponential fall-off up to w_k, so that $w_k = \gamma^k$ could be chosen, assuming $\gamma < 1$. Note that $w_0 = 1$ can be fixed without loss of generality. In order to obtain the best performance of the tracking algorithm as a whole, it may be experimented with different values of k and γ. Thus, the estimate for C_t may be written as:

$$C_t = \frac{\sum_{i=0\ldots k} \gamma^i \Delta D_{t-i} \Delta d_{t-i}}{\sum_{i=0\ldots k} \gamma^i \Delta d_{t-i}^2}$$

5.2.   Implementing the Correction Factor Algorithm

The estimate of C_t needs to be updated each second, where the odometer reading is reliable and the GPS position is available. Providing the GPS is available for

k seconds, it may be worked with an iteratively updated 'moving window' estimate. Now, C_t may be written:

$$C_t = \frac{sum1_t}{sum2_t}$$

where:

$$\begin{cases} sum1_t = \sum_{i=0...k} \gamma^i \Delta D_{t-i} \Delta d_{t-i} \\ sum2_t = \sum_{i=0...k} \gamma^i \Delta d_{t-i}^2 \end{cases}$$

If a record of the last k values of Δd_t and ΔD_t is available, $sum1$ and $sum2$ may be updated at each second, as well as the estimate C_t. This can be seen in:

$$sum1_t = \gamma sum1_{t-1} + \Delta d_t \Delta D_t - \gamma^{k+1} \Delta d_{t-k} \Delta D_{t-k}$$
$$sum2_t = \gamma sum2_{t-1} + \Delta d_t^2 - \gamma^{k+1} \Delta d_{t-k}^2$$

Since the algorithm requires only the values of ΔD and Δd at times t and $t - k$ but not those in between, a 'first in first out' (FIFO) of size $k + 1$ is a useful method of handling the information. Unlike the more usual 'last in first out' (LIFO) stack, popping a value from the stack returns the oldest item on the stack rather than the newest. Thus, at each second, the current values of Δd_t and ΔD_t are computed and pushed onto a FIFO stack. For the first k seconds GPS data is online, no values are popped from the stack. However, after k seconds, values of Δd and ΔD are also popped from the stack. Since the oldest values are popped from the stack, and the stack has had items pushed on to it for k seconds, it implies that Δd_{t-k} and ΔD_{t-k} will be popped.

6.   Calibration if GPS Data Is Recently Online

In the previous section, the method for estimating C_t is used when GPS data is available. If the GPS data is offline, the calibration of C_t cannot take place. The previous section assumed that GPS data is online for a sufficient period of time, so that at least k observations are pushed onto the stack, as well as the supplied GPS positions 'settled' and are reliable. Thus, there is a 'run-in' period of l seconds, where the methods in the previous section cannot be applied. Clearly, l cannot be less than k, but a much larger value may be required. This is to be determined by experiment.

How C_t should be estimated in this l-second time interval, and how the position should be estimated? One possibility is to work initially with a global estimate of C to provide the odometer-based estimate. As time during the burn-in period passes, the estimate should 'drift'. This can be done by combining C_t and C using a weighted average, with the weighting gradually favouring C_t rather than C. The trial method here uses the formula:

$$C_t' = \rho^i C + (1 - \rho^i) C_t$$

where j is the time into the run-in period, and C_t' is the 'combined' estimate of the correction factor, assuming $0 < \rho < 1$.

Finally, sum_1 and sum_2 need to be 'rebuilt' in the first k seconds of this period. That is, for times up to k seconds, the current ΔD_t and Δd_t need to be included into the running mean computation, but ΔD_{t-k} and Δd_{t-k} are not being dropped. Here, it may be written as:

$$\begin{cases} sum1_t = \gamma sum1_{t-1} + \Delta d_t \Delta D_t \\ sum2_t = \gamma sum2_{t-1} + \Delta d_t^2 \end{cases}$$

The estimate of the global C is updated on an ongoing basis when GPS and odometer data are available. It can be noted that:

$$C = \frac{gsum1_t}{gsum2_t}$$

where $gsum1_t$ and $gsum2_t$ may be updated each second, since:

$$\begin{cases} gsum1_t = gsum1_{t-1} + \Delta D_t \Delta d_t \\ gsum2_t = gsum2_{t-1} + \Delta d_t^2 \end{cases}$$

It may be sensible to scale $gsum1_t$ and $gsum2_t$ to avoid rounding errors, for example, we could use a running mean computation such as:

$$\begin{cases} gsum1_t = \dfrac{n-1}{n} gsum1_{t-1} + \dfrac{1}{n} \Delta D_t \Delta d_t \\ gsum2_t = \dfrac{n-1}{n} gsum2_{t-1} + \dfrac{1}{n} \Delta d_t^2 \end{cases}$$

At any time both sums are reduced by a factor n, where n is the number of times the updating algorithm is called. This has no effect on the estimate of C as the numerator and denominator are scaled by the same factor, but it stops them from becoming very large, leading to overflow errors.

7. Putting It All Together

The above algorithms are used on an event-driven basis. Each time a new set of observations is provided, one of three conditions applies:

- GPS and odometer data both available, not during run-in period.
- GPS and odometer data both available, during run-in period.
- Odometer data only available.

The set of actions to be taken in each case are outlined in the algorithms listed in the Appendix.

8.   Alterations to the Correction Factor Algorithm

After the algorithms, described in the previous sections, were implemented, experiments took place in which the algorithms were applied to test data. On the basis of this a number of adjustments were made. In particular, it was found that on occasions the GPS signal was only available for very short periods of time. This occurred either due to a lack of positions provided by the GPS, or because the GPS position provided was rejected as unreliable. As a result of this, there were situations when a GPS location was available for 2 or 3 seconds, then unavailable for a similar length of time, and so on. This led to a problem with algorithm 6 in the Appendix A. Essentially, each time the GPS signal was available the stack was reset, as were *sum*1, *sum*2 and *j*, effectively 'forgetting' the value of C_t prior to GPS signal cut-out. An associated difficulty was that, if the run-in period was longer than 2 or 3 seconds, C_t was never being properly updated during these periods of intermittent GPS availability.

This led to a problem, since either the run-in period had to be very short, as did the size of the stack, or very out-of-date values of C_t were used. Neither option was acceptable. In the first instance, C_t was under-smoothed, leading to erratic odometer based estimates (i.e. a lack of precision) – in the second case, C_t does not vary wildly but is biased, since much of the time values closer to the global C would be supplied – this led to a lack of accuracy.

To overcome this problem, it was noted that GPS cut-outs were never very long, so that calibration of C_t prior to the cut-out would still provide useful information. To this end, algorithm 6 was modified so that the resetting of the stack did not occur when a GPS location became available. The modified algorithm is listed in the Appendix. Experimentation indicated better performance. For the training data, a run in a period of 10 seconds was found, as well as a stack size of 10 – together with $\gamma = 0.8$ the best performance was achieved, in terms of offset error to beacons (for detailed discussion of the methodology see Section 11). It was also found at a later stage that a policy of always returning the odometer-based location estimate (even when GPS was available) improved the accuracy of predictions – this is also reflected in the modified version of algorithm 6.

9.   Height Aiding

In OMMGPS height aiding is used throughout to add an extra equation in the least squares approximation computation of GPS position. Height information obtained, using bilinear interpolation, from a digital terrain model (DTM) is used to achieve 3D GPS point positions, when only three GPS satellites are visible to the receiver.

More importantly, height aiding improves the accuracy of GPS point positions with poor satellite geometry (high PDOD), and when severe signal multipathing occurs (multiple reflected GPS satellite signals). The number of GPS positions

Table 8.1 *Number of GPS positions computed*

Trip no.	GPS only	GPS + Height Aiding
3.4	3834	3834
4	3468	3635
6	3320	3991
9.2	3713	4014

computed is nearly always increased by using additional height information, displayed in Table 8.1. Trip 3.4 is the exception.

10. Implementation

OMMGPS consists of a Dynamic Link Library (DLL), written in C++, together with a GUI for use in ESRI's GIS products ArcView or ArcGIS. The GUI was originally written in ArcView's Avenue and has recently been translated to Visual Basic for use in ArcGIS [Steup and Taylor, 2003]. The GIS is used to visualize the results graphically, using background OS mapping.

11. Data Processing and Results

A number of separate trips along a bus route in central London (Baker Street, Oxford Street, etc.) were made to test the method. During each of these trips, GPS observations using low cost GPS L1 receivers and odometer observations using existing mechanical odometers were taken at each second, on the bus. In total, over 15,000 vehicle positions were computed using OMMGPS. Also, on each trip, the bus recorded the time when it detected a beacon at the beacon's known location, actually to a normal intersect with a five metre offset centre-line, see Figure 8.2.

The positions of the bus at these times were used as the 'true' position of the bus. There are altogether 13 beacons on the bus route, which are used for determining OMMGPS accuracy. Using this data the exact equivalent OMMGPS positions used to calculate distances were obtained by interpolation, applying the OMMGPS positions at the nearest second before and after the beacon detection time. This is simple linear interpolation.

A bus position was available at each second, either computed by GPS or odometer observations. For the whole route, odometer positions are used much more than GPS positions: 70% odometer positions, 30% GPS positions. For the calculation of OMMGPS, position at beacon detection time for the 13 beacons used; four positions were calculated using only GPS, and nine positions were calculated using only odometer. The results obtained for trip 4 are presented in Table 8.2. This bus trip is the one on which the algorithm was developed and

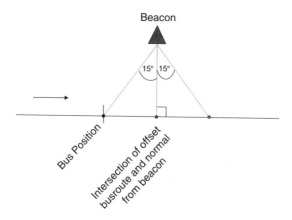

Figure 8.2   *Beacon detection*

Table 8.2   *OMMGPS statistics for all beacons – for 95% – trip 4*

	Error for 100%	Error for 95%
Mean	8.8 m	5.2 m
Standard Deviation	6.6 m	3.6 m
Range	18.7 m	11.2 m
Minimum	0.9 m	0.9 m
Maximum	19.6 m	12.2 m

Table 8.3   *Average errors for all trips*

Trip no	GPS only	OMMGPS	95% cut off OMMGPS
3.4	27.9 m	11.3 m	6.6 m
4	53.7 m	8.8 m	5.2 m
6	>100 m	28.3 m	14.1 m
9.2	40.7 m	22.8 m	14.4 m

tuned, and it shows the excellent potential of the method. The results of the other three trips processed, i.e. trip 3.4, 9.2 and 6, are shown in Table 8.3.

There is a substantial improvement in the accuracy of bus position using OMMGPS instead of only raw GPS. In the case of trip 4, OMMGPS provides a mean error of 8.8 metres compared with 53.7 metres for raw GPS without odometer.

12.   Conclusion

A new algorithm that integrates odometer observations with the existing MMGPS map-matching software was developed and successfully implemented. This new algorithm, called OMMGPS, utilizes map-matching criteria, not to determine which road a vehicle is on, but to determine GPS position precision, in order to calibrate an odometer with the help of GPS. In OMMGPS, GPS pseudorange observations are combined with odometer positions and DTM height information to provide a vehicle position at 1-second epochs. Generally, odometer positioning is used much more often to position the vehicle than GPS. Typically, the ratio is 7:3 odometer positions to GPS positions. This predominant use of odometer positioning is due either to GPS not being available or GPS positions being considered to be too inaccurate to use. This lack of GPS positions is due to satellite masking by buildings or the result of severe GPS signal multipath in the urban canyons of central London.

Four bus trips along the same bus route were used to test OMMGPS. The results obtained from these four trips are most encouraging. In total, over 15,000 vehicle positions were computed using OMMGPS. The positions provided by OMMGPS at the time of beacon detection can be considered to be a random sample of the accuracy provided by OMMGPS, compared to the accuracy provided by GPS alone, that is, if a GPS position was available at all, on or near the beacon detection time. The average error of OMMGPS positions, over all vehicle positions, using the random sample of beacon detection times, is 17.8 metres overall, and 10.1 metres for a 95% cut-off. This compares with an average error for GPS alone of at least 55.6 metres overall.

Moreover, the effectiveness of OMMGPS is entirely dependent on the frequency and accuracy of GPS-derived positions. The GPS data provided for testing compared very poorly with similar raw L1 pseudorange GPS data collected independently along the same bus routes, albeit using a much more expensive receiver and antenna for the independent test observations. Similarly, receiver coordinates collected using another low cost L1 GPS receiver also provided improved positions, although this was most probably due to smoothing provided by this particular receiver's own navigation filter.

In conclusion, the technique developed in OMMGPS works well, and can be further improved with more superior low cost GPS receiver technology or a more careful attention to its operational application. Due to the fact that the method is based on known routes, it is not only appropriate for bus positioning but also for use on the railways.

Appendix: Algorithms

Algorithm 1 Estimate Location of Bus from Odometer Signal

input: float xlast, ylast, d, xroute[1:n], yroute[1:n]; integer position
currentx ← xlast
currenty ← ylast
nextx ← xroute[position]
nexty ← yroute[position]
D1 ← 0
D2 ← distance3d(currentx,currenty,nextx,nexty)
position ← position + 1
currentx ← nextx
currenty ← nexty
nextx ← xroute[position]
nexty ← yroute[position]
D1 ← D2
D2 ← D1 + distance3d(currentx,currenty,nextx,nexty)
beta ← (d − D1) ÷ (D2 − D1)
xestimate ← (1 − beta) × currentx + beta × nextx
yestimate ← (1 − beta) × currenty + beta × nexty
output: position, xlocation, ylocation

Algorithm 2 'Distance3d' Function Used by Algorithm 4

This assumes the existance of a function $DTMZ(x,y)$, which interpolates the z-coordinate of a location (x,y) from a DTM.

input: float x1,x2,y1,y2
z1 ← DTMZ(x1,y1)
z2 ← DTMZ(x2,y2)
d {(x1 − x2)2 + (y1 − y2)2 + (z1 − z2)2}
output: d

Algorithm 3 Update the Value of C_t

input: integer k; float gamma, sum1, sum2, delta.d0, delta.D0, delta.dk, delta.Dk
sum1 ← gamma × sum1
sum1 ← sum1 + delta.d0 × delta.D0
sum1 ← sum1 − power(gamma,k + 1) × delta.dk × delta.Dk
sum2 ← gamma × sum2
sum2 ← sum2 + delta.d0 × delta.d0
sum2 ← sum2 − power(gamma,k + 1) × delta.dk × delta.dk Ct ← sum1 ÷ sum2
output: Ct, sum1, sum2

Algorithm 4 Update the Value of C

input: integer n; float gamma, gsum1, gsum2, delta.d0, delta.D0
factor ← (n − 1) ÷ n
gsum1 ← gsum1 × factor + delta.d0 × delta.D0 × (1 − factor)
gsum2 ← gsum2 × factor + delta.d0 × delta.d0 × (1 − factor)
C ← gsum1 ÷ gsum2 n ← n + 1
output: C, gsum1, gsum2, n

Algorithm 5 Combine C

input: integer j; float rho, Ct, C
output: (1 − power(rho,j)) × Ct + power(rho,j) × C

Algorithm 6 Overview of Events

initialize: Set n, sum1, sum2, gsum1, gsum2 to 0
loop
Repeat each reading
if GPS and odometer available, not during run-in **then**
    Compute delta.D0 and delta.d0
    Push delta.D0 and delta.d0 onto FIFO stack
    Pop delta.Dk and delta.dk from FIFO stack
    Update estimate of Ct using algorithm 4
    Update estimate of C using algorithm 4
    Return GPS location estimate as current position
end if
if GPS and odometer available, during run-in **then**
  if GPS just became available **then**
    set j, sum1, sum2 to 0
    Reset FIFO Stack

end if
Obtain current value of C
if j < k **then**
> Return odometer-based location estimate (algorithm 4) with correction factor C.
> Compute delta.D0 and delta.d0 and push onto FIFO stack
> sum1 ← sum1 × gamma + delta.D0 × delta.d0
> sum2 ← sum2 × gamma + delta.d0 × delta.d0

else
> Compute delta.D0 and delta.d0 and push onto FIFO stack
> Pop delta.Dk and delta.dk from FIFO stack
> Update estimate of C using algorithm 4
> Update estimate of Ct using algorithm 4
> Obtain estimate of C' using algorithm 4
> Return odometer-based location estimate (algorithm 4) with correction factor C'.

> **end if**
end if
if odometer only available **then**
> Return odometer-based location estimate (algorithm 4) with correction factor of the last C_t before the GPS data went offline.
end if
end loop

Modification of Algorithm 6

initialize: Set n, sum1, sum2, gsum1, gsum2 to 0
loop
Repeat each reading
if GPS and odometer available, not during run-in **then**
> Compute delta.D0 and delta.d0
> Push delta.D0 and delta.d0 onto FIFO stack
> Pop delta.Dk and delta.dk from FIFO stack
> Update estimate of Ct using algorithm 4
> Update estimate of C using algorithm 4
> Return odometer-based location estimate as current position
end if
if GPS and odometer available, during run-in **then**
> Obtain current value of C
> **if** j < k **then**
>> Return odometer-based location estimate (algorithm 4) with correction factor C.
>> Compute delta.D0 and delta.d0 and push onto FIFO stack

sum1 ← sum1 × gamma + delta.D0 × delta.d0
sum2 ← sum2 × gamma + delta.d0 × delta.d0
else
   Compute delta.D0 and delta.d0 and push onto FIFO stack
   Pop delta.Dk and delta.dk from FIFO stack
   Update estimate of C using algorithm 4
   Update estimate of Ct using algorithm 4
   Obtain estimate of C′ using algorithm 4
   Return odometer-based location estimate (algorithm 4) with correction
      factor C′.
   end if
end if
if odometer only available **then**
   Return odometer-based location estimate (algorithm 4) with correction
      factor of the last C_t before the GPS data went offline.
end if
end loop

Bibliography

Anai, S. and Ikisu, K., 2002, Intelligent transportation systems by three-dimensional geographic database, in *Proceedings of ION GPS 2000 Conference*, Portland, OR, September. pp. 431–429.

APIC, 2003, http://www.apic.fr, accessed February 2003.

ATI, 2005, Principles and applications of integrating GIS, GPS and remote sensing, http://www.aticourses.com/geomatics.htm, accessed June 2005.

Autodesk, 2003, Discussion groups, http://discussion.autodesk.com/thread.jspa?threaded=222315, accessed June 2005.

Bell, J.W., Amelung, F., Ramelli, A.R. and Blewitt, G., 2002, Land subsidence in Las Vegas, Nevada, 1935–2000: new geodetic data show evolution, revised spatial patterns, and reduced rates, *Environmental and Engineering Geoscience*, 8, 155–174.

Bell, S.B.M., Diaz, B.M., Holroyd, F. and Jackson, M.J., 1983, Spatially referenced methods of processing raster and vector data, *Image and Vision Computing*, 1(4), 211–220.

Bernhardsen, T. 1992, *Geographic Information Systems: An Introduction*, 2nd edn. Chichester: John Wiley & Sons, Inc.

Bernstein, D. and Kornhauser, A., 1998, Map matching for personal navigation assistants. In *Proceedings 77th Annual Meeting, The Transport Research Board*. Washington, DC, pp. 11–15

Berry, J.K., 1987, Fundamental operations in computer-assisted map analysis, *International Journal of Geographical Information Systems*, 1(2), 119–136.

Blewitt, G., 1997, Basics of the GPS technique: observation equations, in B. Johnson (ed.), *Geodetic Applications of GPS*, Nordic Geodetic Commission, Sweden, pp. 10–54

Blewitt, G., Coolbaugh, M., Holt, W., Kreemer, C., Davis, J. and Bennett R., 2003, Targeting of potential geothermal resources in the Great Basin from regional- to basin-scale relationships between geodetic strain and geological structures, *Transactions Geothermal Resources Council*, 27, 3–7.

Blewitt, G. and Taylor, G., 2002, Mapping dilution of precision (MDOP) and map matched GPS, *International Journal of Geographical Information Science*, 16(1), 55–67.

Board, C., 1967, Maps as models, in P. Haggett (ed.), *Models in Geography*, London: Methuen & Co. Ltd, pp. 671–725.

Brown, R.G., 1996, Receiver autonomous integrity monitoring, *Global Positioning System: Theory and Applications*. 2 (A96-20837 04-17), Washington, DC, American Institute of Aeronautics and Astronautics, Inc. (Progress in Astronautics and Aeronautics. Vol. 164), pp. 143–165.

Bullock, B., Geier, G.J., King, T.M. and Kennedy, H.L., 1996, Effectiveness of two satellite tracking in urban canyon environments, in *Proceedings of ION GPS-96, Sixth International Technical Meeting of the Satellite division of The Institute of Navigation*. Kansas City, Missouri, pp. 1699–1707.

Burrough, P.A., 1986, *Principles of Geographical Information Systems for Land Resource Assessment*, Oxford: Oxford University Press.

Burrough, P.A. and McDonnell, A., 1998, *Principles of Geographical Information System: Spatial Information Systems and Geostatistics*, Oxford: Oxford University Press.

Car, A., 1997, *Hierarchical Spatial Reasoning: Theoretical Consideration and its Application to Modelling Wayfinding*, ed. A.U. Frank and P. Haunold, Vienna: Geoinfo Series.

Cardiff City Council, 1999, Real-time passenger information and bus priority system, http://www.cardiff.gov.uk/traffic/internet/telematics/pages_1/ REALTIME%20PASSENGER%20INFORMATION.htm, accessed July 2005.

Carstensen Jr, L.W., 1998, GPS and GIS: enhanced accuracy in map matching through effective filtering of autonomous GPS points, *Cartography and Geographical Information Systems*, 25(1), 51–62.

Carter, J.R., 1988, Digital representations of topographic surfaces, *Photogrammetric Engineering & Remote Sensing*, 54(11), 1577–1580.

Chiang, K., Noureldin, A. and El-Shiemy, N., 2002, Multi-sensor integration using neuron computing for land-vehicle navigation, *GPS Solutions*, 6(4), 209–218.

Chou, Y.H., Rudd, E.I. and Pennington, J., 2005, Emergency 911: integrated GPS/GIS to the rescue, *Geoworld*, http://www.geoplace.com/gw/1998/0898/898emer.asp, accessed June 2005.

Codd, E.F., 1970, A relational model of data for large shared data banks, *Communications of the ACM*, 13(6), 377–387.

Collier, C., 1990, In-vehicle route guidance systems using map matched dead reckoning, in *Proceedings IEEE Position and Navigation Symposium*, Las Vegas, Nevada, pp. 359–363.

Cross, P.A., (1994), Working Paper No. 6: Advanced least squares applied to position fixing, University of East London.

Date, C.J., 2000, *An Introduction to Database Systems*, 7th edn, Reading, MA: Addison-Wesley.

Davis, Jr., C.A. and Albuquerque de Vasconcelos Borges, K., 1994, Object-oriented GIS in practice – URISA 1994, in *Annual Conference Proceedings*, Washington, DC: Urban and Regional Information Systems Association, pp. 1786–1795.

Demers, M.N., 2000, *Fundamentals of Geographic Information Systems*, 2nd edn, Chichester: John Wiley & Sons. Inc.

Department of Defense and Department of Transportation, 1992, *Federal Radionavigation Plan*, Washington, DC: USA Government.

DOE, 1987, *Handling Geographic Information*, Report of the Committee of Enquiry chaired by Lord Chorley, London: HMSO.

Dorey, M., 2002, Digital elevation models for intervisibility analysis and visual impact assessment, PhD dissertation, University of Glamorgan, Pontypridd, Wales.

DOT, 2000, Social exclusion and the provision of public transport, http://www.dft.gov.uk/ stellent/groups/dft_mobility/documents/page/dft_mobility_506795.hcsp, 27 October 2000, accessed July 2005.

English, B.C., Roberts, R.K. and Sleigh, D.E., 2000, Spatial distribution of precision farming technologies in Tennessee, Research Report 00–08, Department of Agricultural Economics and Rural Sociology, Tennessee Agricultural Experiment Station, the University of Tennessee.

ESRI, 2003, http://www.esri.com, accessed February 2003.

Fairbairn, D. and Taylor, G., 2002, Data collection issues in virtual reality for urban geographical representation and modelling, in P. Fisher and D. Unwin (eds), *Virtual Reality in Geography*, London: Taylor and Francis, pp. 220–238.

Fei, P., Qishan, Z. and Zhongkan, L., 2000, The application of map matching method in GPS/INS integrated navigation system, International Telemetering Conference, USA Instrument Society of America, 36(2), 728–736.

Fisher, P. and Unwin, D. (eds), 2001, *Virtual Reality in Geography*, London: Taylor and Francis, pp. 211–219.

French, R.L., 1997, Land vehicle navigation and tracking, *Global Positioning System: Theory and Applications*, Vol. II: 275–301.

Gahegan, M.N., 1989, An efficient use of quadtrees in a geographical information system, *International Journal of Geographical Information Systems*, 3(3), 201–214.

Garten, M., 2003, Data integration issues for a farm GIS-based spatial decision support system, MPhil thesis, University of Glamorgan, Wales.

Gartner, G., Radoczky, V. and Retscher, G., 2005, Location technologies for pedestrian navigation, The Geospatial Recourse Portal, http://www.gisdevelopment.net/magazine/years/2005/apr/location.htm, accessed June 2005.

Ge, L., Chang, H., Janssen, V. and Rizos, C., 2002, Integration of GPS, radar interferometry and GIS for ground deformation monitoring, http://www.gmat.unsw.edu.au/snap/publications/ge_etal2003c.pdf, accessed June 2005.

GE Energy, 2004, Smallworld 4 product suite, http://www.gepower.com/prod_serv/products/gis_software/en/smallworld4.htm, accessed September 2004.

Goodchild, M. and Gapal, S., 1988, *Accuracy of Spatial Databases*, London: Taylor & Francis.

Gray, P., Kulkarni, M.D., Krishnarao, G. and Paton, N.W., 1992, *Object-Oriented Databases: A Semantic Data Model Approach*, Englewood Cliffs, NJ: Prentice & Hall International.

Grenzdorffer, G., 2000, GIS for precision farming, *GIM International*, 14(8), 12–15.

Groves, P.D. and Handley, R.J., 2004, Optimising the integration of terrain referenced navigation with INS and GPS, *Proceedings ION GNSS 17th International Technical Meeting of the Satellite Division*, 21–24 Sept. 2004, Long Beach, CA, pp. 1048–1059.

Hailes, T.A., 1999, Integrating technologies: DGPS dead reckoning and map matching. *International Archives of Photogrammetry and Remote Sensing*, 32(2W1), 1.5.1–1.5.8.

Harley, J.B., 1975, *Ordnance Survey Maps: A Descriptive Manual*, London: HMSO.

Herring, J.R., 1987, Topologically integrated geographic information systems, *Proceedings Autocarto 8*, pp. 282–291.

Heslop, D. and Taylor, G., 1998, City of Newcastle upon Tyne Urban Archaeological Geographical Information System (GIS), Computer Applications in Archaeology, Third United Kingdom Conference, University of Southampton.

IEEE-VR2002, 2002, Tenth Symposium on Haptic Interfaces for Virtual Environment and Teleoperator Systems, http://www.rcl.ece.ubc.ca/events/haptics-symposium/index.html, accessed February 2003.

Information Technology Education, 2005, National GIS/GPS Integration Team, http://www.csrees.usda.gov/nea/technology/in_focus/infotech_if_gisgps.html.

Interagency GPS Executive Board, 2000, *Frequently Asked Questions About SA Termination*, http://www.igeb.gov/sa/faq.shtml, accessed 12 December 2000.

Jones, C., 1997, *Geographical Information Systems and Computer Cartography*, Harlow: Longman.

Kaplan, E.D., 1996, Performance of standalone GPS, in *Understanding GPS: Principles and Application*, Boston: Artech House Publications, pp. 306–311.

Kealy, N., Tsakiri, M. and Stewart, M., 1999: Land vehicle navigation in the urban canyon: a Kalman filter solution using integrated GPS, GLONASS and dead reckoning, in *Proceedings of ION GPS 99 Conference*, pp. 509–518.

Kidner, D., Dorey, M., Smith, D., 1999. What's the point? Interpolation and extrapolation with a regular grid DEM, in Proceedings of GeoComputation'99, Virginia, http://www. geovista.psu.edu/sites/geocomp99/Gc99/082/gc_082.htm, accessed July 2002.

Kidner, D.B., Higgs, G. and White, S.D. 2002, *Socio-economic Applications in Geographical Information Science: Innovations in GIS 9*, London: Taylor and Francis.

Kidner, D.B., 2003. Higher order interpolation of regular grid digital elevation models, *International Journal of Remote Sensing*, 24(14), 2981–2987.

Kim, J.S., Lee, J.H., Kang, T.H., Lee, W.Y. and Kim, Y.G., 1996, Node based map matching algorithm for car navigation system, in *Proceedings of the 29th ISATA Symposium*, Florence, 10, 121–126.

King, R., 2004, *State Plane Coordinate System*, http://gislounge.com/features/aa032700. shtml, last accessed September 2004.

Lake, R., Burggraf, D., Trninic, M. and Rae, L., 2004, *Geography Mark-Up Language: Foundation for the Geo-Web*, New York: John Wiley & Sons, Ltd.

Lee, J. and Kwan, M.P., 2000, A 3-D object-oriented data model for representing geographic entities in built-environments, paper presented in 96th AAG at Pitts., PA, USA.

Leica, 1999. *Getting Started with Static and Kinematic Surveys*, Heerbrugg, Switzerland: Leica Geosystems AG.

Levy, L.J., 1997, The Kalman filter: navigation's integration workhorse, *GPS World*, 8(9), 65–71.

Li, J., Taylor, G. and Kidner, D., 2003, Accuracy and reliability of map matched GPS coordinates: dependence on terrain model resolution and interpolation algorithm, in *Proceedings of AGILE Conference on Geographic Information Science*, Lyon, France.

Löhnert, E., Mundle, H., Wittmann, E. and Heinrichs, G., 2003, PARAMOUNT – experiences and results of a LBS prototype for mountaineers, in *Proceedings ION GPS/GNSS 2003*, 9–12 September 2003, Portland, OR, pp. 1620–1627.

Longley, P.A., Goodchild, M.F., Maguire, D.J. and Rhind, D.W., 2001, *Geographic Information Systems and Science*, Chichester: John Wiley & Sons, Ltd.

Mackett, R. and Titheridge, H., 2004, A methodology for the incorporation of social inclusion into transport policy, paper presented at the World Conference on Transport Research, 4–8 July 2004, Istanbul, Turkey.

Mallet, P. and Aubry, P., 1995, A low-cost localisation system based on map matching technique, in *Proceedings of the International Conference on Intelligent Autonomous Systems*, Karlsruhe, Germany, pp. 72–77.

Martin, D, 1996, *Geographic Information Systems: Socioeconomic Applications*, 2nd edn, London: Routledge.

Martin, E., 2000, *Geographic Information Systems Analysis Lecture 27*, Universityb of Washington, http://faculty.washington.edu/geog460/lectures/lec27.html, accessed February 2003.

Matt, M., 2004, Editorial / April 2004 / Finding your friends with SMS and GPS, http://www.mtekk.com.au/browse/page766.html, accessed October 2005.

Mattos, P.G., 1993, Intelligent sensor integration algorithms for vehicles, in *Proceedings of ION GPS-93, Sixth International Technical Meeting of the Satellite Division of The Institute of Navigation*, Salt Lake City, Utah, pp. 1591–1597.

Mayhew, D. and Kachroo, P., 1998, Multi-rate sensor fusion for GPS using Kalman filtering, fuzzy methods and map-matching, in *Proceedings of SPIE Conference on Sensing and Controls with Intelligent Transportation Systems*, Boston, USA, November 1998, vol. 3525, pp. 440–449.

McCall, R.B., 1970, *Fundamental Statistics for Psychology*. New York: Harcourt, Brace & World, Inc.

Mezentsev, O., Lu, Y., Lachapelle, G. and Klukas, R., 2002, Vehicle navigation in urban canyons using a high sensitivity GPS receiver augmented with a low cost rate gyro, in *Proceedings of ION GPS 2000 Conference*, Portland, OR, USA, September 2002, pp. 363–369.

Mountain, D. and Raper, J., 2001, Positioning techniques for location-based services (LBS): characteristics and limitations of proposed solutions, *Aslib Proceedings: New Information Perspectives*, 53(10), 404–412.

National GIS/GPS Integration Team (2005), http://www.tnstate.edu/iager/gisgps/main. htm.

NSF Panel on Graphics, Image Processing, and Workstations, 1987, *Visualization in Scientific Computing*, NSF, Virginia, USA.

Ochieng, W.Y., Quddus, M.A. and Noland, R.B., 2004, Integrated positioning algorithms for transport telematics applications, in *Proceedings ION GNSS 17th International Technical Meeting of the Satellite Division*, Sept. 2004, Long Beach, CA, pp. 692–705.

OGC, 2004, Welcome to OGC, http://www.opengeospatial.org, accessed October 2004.

Ordnance Survey, 1999, *Ordnance Survey Datum Transformation OSTN97 and Geoid Model OSGM91*, Southampton: OS.

Ordnance Survey, 2001, *Land-Form PROFILE User Guide v4.0*, Southampton: Ordnance Survey.

Ordnance Survey, 2004, Interactive Guide to the National GridOrdnance Survey, http://www.ordnancesurvey.co.uk/oswebsite/freefun/nationalgrid/nghelp1.html, accessed September 2004.

Parker, D., 1990, Land information databases, in T.J.M. Kennie (ed.), *Engineering Surveying Technology*, Glasgow: Blackie, pp. 427–477.

Petrovello, M.G., Cannon, M.E. and Lachapelle, G., 2003: Quantifying improvements from the integration of GPS and a tactile grade INS in high accuracy navigation systems, in *Proceedings of ION NTM Conference*, Anaheim, CA, January 2003.

Peuquet, D.J., 1984, A conceptual framework and comparison of spatial data models, *Cartographica*, 21(4), 66–113.

Pietilä, S. and Williams, M., 2002, Mobile location applications and enabling technologies, in *Proceedings ION GPS 2002*, 24–27 September 2002, Portland, OR, pp. 385–395.

Piplapure, A., 2004, A practical and economic scheme for implementing integrated GIS/network analysis system at the Indian electricity distribution utilities, http://www.gisdevelopment.net/application/utility/power/utilityp0014b.htm, accessed September 2004.

Poiker, T.K. and Crain, I.K., 1986, Geographic Information Systems, in M. Nurtig (ed.), *The Canadian Encyclopedia*, http://www.sfu.ca/geography/people/faculty_emeritus/ emeritus_sites/TomPoiker/publications.htm.

Rajé, F., 2003, The impact of transport on social exclusion processes with specific emphasis on road user charging, *Transport Policy*, 10(4), 321–338.

Ramjattan, A. and Cross, P.A., 1995, A Kalman filter model for an integrated land vehicle navigation system, *Journal of Navigation*, 49(2), 293–302.

Raper, J., 2000, *Multidimensional Geographic Information Science*, London: Taylor & Francis.

Russell, W.S., 1995. Polynomial interpolation schemes for internal derivative distributions on structured grids, *Applied Numerical Mathematics*, 17, 129–171.

Schiller, J. and Voisard, A. (eds), 2004, *Location-Based Services*, San Francisco: Elsevier Science & Technology Books.

Scott, C., 1994, Improving GPS positioning for motor-vehicle through map matching, in *Proceedings of ION GPS-94, Seventh International Technical Meeting of the Satellite Division of The Institute of Navigation*, Salt Lake City, Utah, pp. 1391–1140.

Sickle, J.V., 1996. *GPS for Land Surveyors*, Chelsea, MI: Ann Arbor Press, Inc.

Singer, M.H., 1993, A general approach to moment calculations for polygons and line segments, *Pattern Recognition*, 26(7), 1019–1028.

Smallworld, 2003, http://www.gepower.com/dhtml/network_solutions/en_us/smallworldtechnology.

Social Exclusion Unit, 2003, *Making the Connections: Final Report on Transport and Social Exclusion*, London: Office of the Deputy Prime Minister.

Southern, R., 2002, Mobile location based services, in *Proceedings AGI 2002 Conference*, London.

Steede-Terry, K., 2000, *Integrating GIS and the Global Positioning System*, Redlands, CA: ESRI Press.

Stephen, J. and Lachapelle, G., 2000, Development of a GNSS-based multi-sensor vehicle navigation system, in *Proceedings of ION NTM Conference*, Anaheim, CA, USA, January 2000.

Steup, D. and Taylor, G., 2003, Porting GIS applications between software environments, in *Proceeding of the GIS Research UK 2003 11th Annual Conference*, London, pp. 166–171.

Syed, S. and Cannon, M.E., 2004, Three dimensional fuzzy logic based-map matching algorithm for location based service applications in urban canyons, in *Proceedings ION GNSS 17th International Technical Meeting of the Satellite Division*, 21–24 Sept. 2004, Long Beach, CA, pp. 2404–2415.

Tanaka, J., Hirano, K., Itoh, T., Nobuta, H. and Tsunoda, S., 1990, Navigation system with map-matching method, in *Proceedings of the SAE International Congress and Exposition*, Detroit, pp. 45–50.

Taylor, G., Multimedia and Virtual Reality, 1999, NCGIA Core Curriculum in Geographic Information Science, http://www.ncgia.ucsb.edu/giscc/units/u131/u131_f.html.

Taylor, G. and Blewitt, G., 1999, Virtual differential GPS and road reduction filtering by map matching, in *Proceedings of ION'99, Twelfth International Technical Meeting of the Satellite Division of the Institute of Navigation*, Nashville, TN, USA, pp. 1675–1684.

Taylor, G. and Blewitt, G., 2000, Road reduction filtering using GPS, in *Proceedings of 3rd AGILE Conference on Geographic Information Science*, Helsinki, Finland, pp. 114–120.

Taylor, G., Blewitt, G., Steup, D., Corbett, S. and Car, A., 2001a, Road reduction filtering for GPS-GIS navigation, *Transactions in GIS*, 5(3), 193–207.

Taylor, G. and Heslop, D., 1998, The use of Geographical Information Systems (GIS) and visualization technology for the management of archaeological records, in *Proceedings: Computer Applications in Archaeology*, Barcelona, Spain.

Taylor, G., Uff, J. and AI-Hamadani, A., 2001b, GPS positioning using map-matching algorithms, drive restriction information and road network connectivity, in *Proceedings of the GIS Research UK 9th Annual Conference*, Pontypridd, Wales, pp. 114–119.

Thurston, J., Poiker, T.F. and Moore, J.P., 2003, *Integrated Geospatial Technologies: A Guide to GPS, GIS and Data Logging*, Chichester: John Wiley & Sons, Ltd.

Tomlinson, R.F., 1984, Geographic information systems: a new frontier, *The Operational Geographer*, 5, 31–35.

Trimble, 2005, Trimble Pathfinder GPS Mapping Systems, http://www.ascscientific.com/pthfindr.html, accessed June 2005.

Udani, P.M. and Goel, R.K., 2002, GPS enabled mobile GIS services, http://www.gisdevelopment.net/technology/gps/techgp0021a.htm, accessed June 2005.

Usery, E.L., Pocknee, S. and Boydell, B., 1995., Precision farming data management using geographic information systems, *Photogrammetric Engineering and Remote Sensing*, 61, 1383–1391.

Wasinger, R., Stahl, C. and Krüger, A., 2004, M3I in a pedestrian navigation and exploration system, http://w5.cs.uni-sb.de/~stahl/m3i/m3i-2003MobileHCI.pdf, accessed June 2005.

Weibel, R. and Heller, M., 1991, Digital terrain modeling, in D.J. Maguire, M.F. Goodchild and D.W. Rhind (eds), *Geographical Information Systems: Principles and Applications*, London: Longman, vol. 1, pp. 269–297.

Welsh Consumer Council, 2001, *Bus Travel in Wales: A Consumer's Journey*, Cardiff: Welsh Consumer Council.

Wescott, K. and Brandon, R., 1999, *Practical Applications of GIS for Archaeologists*, London: Taylor & Francis.

White, C.E., Bernstein, D. and Kornhauser, A.L., 2000, Some map matching algorithms for personal navigation assistants, *Transportation Research Part C: Emerging Technologies*, 8(1–6), 91–108.

White, M., 1991. Car navigation systems, in D.J. Maguire, M.F. Goodchild and D.W. Rhind (eds), *Geographical Information Systems: Principles and Applications*, London: Longman, Vol. 1, pp. 115–125.

Wilson, J.P., 1999, Local, national and global applications of GIS in agriculture, in P. Longley, M. Goodchild, D. Rhind and D. Maguire (eds), *Geographical Information Systems: Principles and Applications*, John Wiley & Sons, Ltd., vol. 2, pp. 981–998.

Wise-McLain, P. and Murphy, M.D., 1993, A GPS/dead reckoning system for tracking and GIS application, in *Proceedings of Proceedings of ION GPS-93, Sixth International Technical Meeting of the Satellite Division of the Institute of Navigation*, Salt Lake City, Utah, pp. 1631–1636.

Index

With kind thanks to Geoffrey Jones of Information Index for compilation of this index.

Learning Resources
Centre